DATE DUE

By Lowell Dingus, with Mark A. Norell and Eugene Gaffney

DISCOVERING DINOSAURS:
Evolution, Extinction, and the Lessons of Prehistory

By Lowell Dingus

NEXT OF KIN:
Great Fossils at the American Museum of Natural History

By Lowell Dingus and Luis M. Chiappe

THE TINIEST GIANTS:
Discovering Dinosaur Eggs

WALKING ON EGGS

The Astonishing Discovery of Thousands
of Dinosaur Eggs in the Badlands of Patagonia

Luis M. Chiappe
and
Lowell Dingus

SCRIBNER
New York   London   Toronto   Sydney   Singapore

SCRIBNER
1230 Avenue of the Americas
New York, NY 10020

DESIGNED BY ERICH HOBBING

Set in Electra

Manufactured in the United States of America

1   3   5   7   9   10   8   6   4   2

Library of Congress Cataloging-in-Publication Data
Chiappe, Luis M.
Walking on eggs : the astonishing discovery of thousands of dinosaur eggs in the badlands of
Patagonia / Luis M. Chiappe and Lowell Dingus.
p. cm.
Includes index.
1. Dinosaurs—Eggs—Argentina—Nequén (Province) 2. Fossils—Argentina—Nequén
(Province) 3. Chiappe, Luis M. 4. Dingus, Lowell. I. Dingus, Lowell. II. Title.
QE861.6.E35 C48 2001
567.9'0982'72—dc21
2001020280
ISBN 0-7432-1211-8

To all the members of our team,
who share with us the thrill for exploration
that made this investigation possible

Contents

List of Illustrations

Photographs following page 160

PREFACE

For many people, few professions conjure up more romantic notions than paleontology. The idea of spending weeks at a time wandering around remote and exotic locales in search of dinosaur fossils sounds much more intriguing than being shackled to a desk in an office. As paleontologists, we can both attest that some of these notions are true. The quest to find fossils is enthralling, and the moment of discovery is exhilarating. Experiencing the vastness of dinosaur-producing landscapes, the cool breeze that blows at dusk, and the starry sky that envelops you at night provides priceless moments and memories. Neither of us would be willing to trade our time in the field for higher-paying jobs that restricted us to working exclusively in an office or laboratory. Yet, life on a paleontological expedition is not always easy. Modern expeditions are enormously complex operations, requiring teamwork, dedication, sacrifice, and diverse scientific specialties.

The reasons for this are simple. First, no single paleontologist possesses all the knowledge and skills required to conduct the array of interdisciplinary investigations necessary to reconstruct a complete picture of a fossil locality and its ancient inhabitants. Over the last century, paleontology has come to rely on contributions from numerous scientific disciplines in unraveling the mysteries of past life on earth. Consequently, paleontological expeditions require a team of specialists to investigate all the questions that must be answered.

Life in the field for such crews is intense. The needs of these specialists, such as equipment and transportation, must be accommodated as best as possible, but limitations of time and money require each individual to make compromises for the benefit of the team. Beyond that, conditions are often harsh; meals are not always cooked to your liking; and basic resources are commonly scarce. Nonetheless, the crew must cooperate with each other for weeks, if not months, and the crew's morale can easily be disrupted by selfish behavior. Although moments of fun and frivolity are an attractive component of fieldwork, long days of hard work are the norm. Conducting the research is your primary obligation.

In modern paleontological research, one can approach the study of fossils from two basic perspectives, biology or geology, but in the end, both must be done for any comprehensive study of a new locality. When one pursues the biological approach, one studies the fossil animal's anatomy to piece together a picture of what the animal looked like, how the animal is related to other animals, and how the animal's body worked. When one pursues the geological approach, one studies the rocks that the fossils are preserved in to piece together a picture of when the animal lived and the kind of environment it inhabited.

When we decided to mount our first, full-scale expedition to Patagonia in the fall of 1997, we knew that one thing we had going for us was that, together, we represented each of these two basic perspectives. Luis had trained as a paleobiologist while a student in his native Argentina. He had studied biological sciences at the University of Buenos Aires in the 1980s and early 1990s, participating in numerous paleontological expeditions under the supervision of one of Argentina's most famous paleontologists, José Bonaparte, whom you will learn more about later. Luis's experience led him to specialize in the study of modern and fossil reptiles, including crocodiles, dinosaurs, and especially ancient birds. Lowell took a completely different career path, which emphasized geological studies at the University of California at Riverside and Berkeley, during the 1970s and early 1980s. His fieldwork took him to both Oregon and Montana, where he studied dinosaur extinction with William Clemens, just at the time that Walter and Luis Alvarez proposed that the impact of an asteroid had caused the extinction of dinosaurs 65 million years ago.

Our two divergent academic paths crossed when we were both working at the American Museum of Natural History in New York. Luis was a postdoctoral researcher working on grants designed to fill the gaps in our knowledge concerning the earliest stages in the evolution of birds. Lowell was directing a $50-million renovation of the museum's famous exhibition halls of dinosaurs and other fossil vertebrates. Yet, we both yearned to continue our work in the field, and that opportunity presented itself when the museum initiated expeditions to Mongolia in 1991 under the supervision of Michael Novacek and Mark Norell in the Department of Vertebrate Paleontology.

The Gobi Desert of southern Mongolia has been a mecca for paleontologists since the 1920s, when crews from the American Museum of Natural History under the direction of explorer Roy Chapman Andrews and paleontologist Walter Granger discovered the first well-preserved clutches of dinosaur eggs at the Flaming Cliffs. In the 1930s, political turmoil resulting from the Communists' rise to power in Mongolia put an end to these Central Asiatic expeditions. From the 1950s to the 1980s, paleontologists from the Soviet Union and Poland continued to conduct expeditions in association with their Mongolian colleagues, but those collaborative efforts once again changed drastically in the late 1980s and early 1990s when the Soviet Union collapsed. In 1990, after discussions with the Mongolian Academy of Sciences, the American Museum of Natural History was asked to renew its paleontological exploration of the Gobi in cooperation with Mongolian paleontologists. Those expeditions began in 1991. Lowell signed on as the expedition's chief geologist, and soon after, Luis joined up to search for fossils of small carnivorous dinosaurs like *Velociraptor* that could shed new light on the evolutionary origin of birds.

Our experiences together during these field seasons were filled with fantastic weeks of adventure and discovery. Luis teamed up with the paleontologists to participate in the discovery and subsequent research concerning numerous new dinosaurs, including the first known embryo of a carnivorous dinosaur and the first known dinosaur to be found brooding eggs on top of its nest. Lowell and the other geologists refined the age of the area's fossil localities by studying the rocks' paleomagnetism and established that most of the fossilized animals had perished under massive avalanches of wet sand, which had flowed down the steep faces of the sand dunes that had studded the ancient desert.

In 1997, we teamed up again to lead a crew of paleontologists to explore a remote desert area of northern Patagonia. Although we didn't know what we might find, we did know that this region had produced many spectacular specimens of dinosaurs and fossil birds over the last century. We also knew that if we pooled our previous field experiences and scientific training, we had the nucleus for a well-rounded scientific team of paleobiologists and geologists.

The discoveries our crew made raised dozens of scientific mysteries. To solve them, we would have to draw upon our own scientific training and enlist the support of numerous experts in a variety of other specialized disciplines throughout paleontology and geology. This book will tell the story of how our crew made those discoveries and how we solved some scientific riddles surrounding a natural catastrophe deeply rooted more than 70 million years in the past.

Acknowledgments

During our four years exploring the badlands around Auca Mahuida, many colleagues, technicians, students, and volunteers joined the ranks of our team. Their names are too numerous to list here, but they all deserve special recognition because without their efforts it would have been impossible to accomplish what we did. You will be introduced to many of them personally in the course of our story. Among them, we would like to extend our deepest gratitude to our friend and colleague Rodolfo Coria, whose company in the field and partnership in our research have been both essential and gratifying. We also wish to thank our literary agents, Edite Kroll and Samuel Fleishman, for their encouragement and help, as well as our editor, Gillian Blake, along with Rachel Sussman and the staff, for their many constructive suggestions. We thank Nicholas Frankfurt for portraying so vividly our inferences about dinosaurs so long extinct.

Our research would not have been possible without the generous financial support from American Honda, the Ann and Gordon Getty Foundation, Fundación Antorchas, the InfoQuest Foundation, the Municipalidad de Plaza Huincul, the National Geographic Society, the Phillip McKenna Foundation, the Secretaria de Cultura de la Provincia del Neuquén. Sponsorship of the project under the auspices of the Natural History Museum of Los Angeles County, the Museo Municipal "Carmen Funes," and the American Museum of Natural History also proved invaluable.

Finally, we are especially grateful for the help provided by Larry and Shammy Dingus, who encouraged us to undertake the project before it began and steadfastly supported our efforts throughout.

WALKING
ON EGGS

The Moment of Discovery

Eggs were everywhere. As we strode across the mud-cracked flats exposed beneath the banded ridges of crimson rock that radiated under the searing Patagonian sun, crew members began kneeling down to examine small, dark gray fragments of rounded rock with a curious texture. Picking up these chunks for closer inspection, we could see that the surface was sculpted into hundreds of small bumps and depressions. We knew immediately from the distinctive texture that we had found something startling—dinosaur eggs.

A quiet but elated sense of amazement enveloped the crew. Our morning routine of casual prospecting had instantly turned into a moment of stunning discovery. With a bit of careful reconnaissance and a healthy dose of good luck, we had stumbled across a fantastic new site, the kind of site we had been hoping to discover our entire lives—remote, untouched, and crammed full of fossils that no one had ever seen before.

We slowly came to our senses and began to take stock. Scanning the scene around us, we were once again stunned by the sight of thousands of egg fragments littered across the desolate Patagonian landscape. In many places the fragments of eggs were so abundant that we couldn't walk without stepping on pieces of fossil eggshell. These were not the small bones of the ancient birds we had originally hoped to find, but serendipity, a common companion on paleontological expeditions, had not let us down.

Before us lay a scene of unparalleled paleontological carnage in the form of thousands of eggs. Beneath the weathered surface of the ground, many eggs appeared to be completely intact, as if unhatched. They were large, as dinosaur eggs go, but not enormous—about five to six inches across and almost spherical. As we stared at these fossil treasures, the eggs seemed to stare back and pose numerous puzzling questions. Luis pondered the paleontological mysteries:

- What kind of dinosaurs had laid the eggs?
- How had they been preserved for millions of years after they were laid?
- Were they laid in discrete nests or deposited randomly across the ancient landscape?
- Were they all laid during the same breeding season?
- What other animals had lived in the area at the time?

Meanwhile, Lowell wondered more about the geological riddles:

- Exactly how many million years ago were the eggs laid?
- What was the environment like around the nesting ground?
- And most intriguingly, what kind of calamity had occurred to keep all these eggs from hatching?

Luis knew that eggs somewhat similar in size and shape, along with even larger ones with thicker shells, had previously been found in other areas of the world, as well as in Patagonia. In fact, the first dinosaur eggs ever found, near the end of the nineteenth century, were similar eggs from layers of rock in France that dated from near the end of the Mesozoic era, about 70 million years ago. Paleontologists had long speculated that these large eggs had been laid by sauropods—a group of titanic, long-necked, four-legged dinosaurs that includes the largest animals ever known to have walked the earth. But no one had ever found fossils of embryos inside those eggs, so the identity of the dinosaurs that had laid the eggs could not be established for certain.

After discussing the puzzles posed by the eggs, our sense of elation quickly dissolved into determination when Luis issued the appropriate challenge. We had to find fossil bones inside one of the eggs to solve the mystery of which kind of dinosaur had laid them. So, the whole crew once again set off across the flats in pursuit of an even rarer fossil treasure: a fossilized embryo.

We knew that finding an embryo would not be easy. The fragile bones and tissues of an embryonic animal almost always decay rapidly after it dies, and these embryos had died more than 65 million years ago when all the large dinosaurs went extinct. To find an embryo in the eggs would require not only unique conditions in the ancient environment where the fossils had formed, but also a lot of patience. We would have to bend over to pick up thousands of egg fragments and carefully examine each one's fossilized contents. It was a daunting task, but we felt that our chances were good because there were so many eggs and the rocks in this region had previously produced unlikely fossil jewels. The expedition had just begun, and the crew's morale could not be higher. We had weeks ahead of us for exploring this paleontological paradise.

CHAPTER ONE

Patagonia

A Faraway Land Full of Fossils

We were by no means the first paleontologists to find fossils in Argentina. In fact, we were simply the latest in a long line of explorers and scientists who had traveled to Patagonia in search of gold and scientific treasures.

Discovered by the Europeans in the early 1500s, Patagonia was soon dubbed an "Island of Giants," even though the region is not separated by water from the rest of the South American continent. This nickname is directly related to the origin of Patagonia's name. Some historians have argued that the word *Patagonia* comes from the current meaning of the Spanish term *patagon*, which means "big foot." As the story goes, explorers commissioned by the crown of Spain described the aboriginal inhabitants they encountered as having big feet, thus their land was called Patagonia. But it seems far more likely that the real origin of the name comes from the great explorer Ferdinand Magellan—who died while leading the first expedition to sail completely around the world. When Magellan first encountered the Tehuelche Indians of southern South America during his voyage in 1520, he called them Patagoni (the Latinized plural of *Patagon*). It appears that Magellan drew this name from a famous tale of chivalry that was popular at the time. In *Primaleon*, a courageous knight decides to fight the giant "Patagon," who lives on a remote island. Obviously, *Primaleon* was based on an even older story that had originated in ancient Greece: *The Odyssey*, in which mighty Odysseus fights the giant, one-eyed Cyclops on a legendary island.

Map of Argentina
showing the location
of Auca Mahuevo,
our nesting site.

An entry in Magellan's diary for May 19, 1520, reads, "Two months passed by without our seeing a single inhabitant of the country. One day, when we least expected it, a huge man appeared before us. He stood on the sand, almost naked, and sang and danced while throwing dust over his head." Due to Magellan's description, Europeans began to think of the "Patagoni" as giants, but in reality these people were of normal height. In all probability, Magellan was simply trying to enhance his own image as a courageous explorer and legendary hero—as if being the leader of the first crew to sail completely around the world wasn't enough.

Today, Patagonia encompasses an enormous region that covers most of the cone at the southern tip of South America. Although the Mercator projection that is commonly used to draw maps may make it look small on a map, Patagonia is nearly half the size of Greenland and larger than the states of Texas and Oklahoma combined, with an area of more than four hundred thousand square miles. Most of Patagonia is contained within Argentina, east of the high and frigid

peaks that make up the jagged backbone of the Andes m̲
range and south of the Río Colorado. The Argentine portio̲
Patagonia is subdivided into five provinces: Neuquén and Río Negr̲
in the north, Chubut and Santa Cruz to the south, and the island of
Tierra del Fuego at the southern tip.

Little rain makes it over the Andes to fall on the hills and plains to
the east. Thus, east of the magnificent forests and lakes that adorn the
ice-capped peaks of the Andes, Patagonia is mostly a barren and
dusty desert, whipped by strong gusts of wind. The sun cracks the surface of the earth during the blazing heat of summertime, but the winter brings bitterly cold temperatures to some parts of the region. In this
desolate and inhospitable land live the puma, the condor, and the
ostrichlike rhea. Beneath it are the abundant remains of past organisms, such as the dinosaur eggs that we found, because Patagonia is
covered by layer upon layer of rocks that entomb fossils of mammals
and dinosaurs that lived millions of years ago.

In fact, Argentina and other regions of South America played an
important role in the development of Charles Darwin's theory of evolution by natural selection. Between 1833 and 1836, the HMS *Beagle*
anchored off the southern shore of Argentina. There, at Bahía Blanca,
Darwin wandered along the beach and collected fossils of Ice Age
mammals. He was amazed at the contrast between the modern fauna
and the fauna of Ice Age Argentina. In a two-hundred-square-yard area
of reddish mud and gravel bluffs along the beach, Darwin and his colleagues discovered fossils of huge mammals. The ancient fauna
included several species of giant ground sloths and a glyptodont—an
armadillo-like animal with a nearly complete covering of bony armor
arranged in small polygonal plates across the body. Some glyptodonts
grew to lengths of five or six feet, resembling small tanks. The giant
sloths, of which Darwin collected a nearly complete skeleton, were also
enormous, reaching heights of over ten feet.

Darwin and other members of the *Beagle*'s crew sailed farther
south to Puerto San Julián, where Darwin found beds of rock containing fossil oysters up to one foot in diameter, as well as shells of
other marine creatures. Later research has shown that these animals
lived more than 20 million years ago, when much of Patagonia was
submerged under a shallow sea. In other rock layers above the oyster-
bearing beds, Darwin found a skull-less skeleton of a terrestrial fossil

Archival illustration of glyptodonts and giant sloths.

mammal. Embarking from the mouth of the Santa Cruz River south of the port of San Julián, the crew took a three-week excursion deep into the interior of Patagonia. The party reached the base of the imposing Andes before retracing their path to the Atlantic coast. Perhaps if Darwin had had more time to explore this region, he may well have discovered that it contains one of the largest known treasures of fossil mammals and dinosaurs in the world, along with Mongolia, China, and the western part of the Great Plains in North America.

The first dinosaur fossils from Patagonia to come to the attention of the scientific community were discovered by an Argentine army officer, Captain Buratovich, near the city of Neuquén in 1882. This region is just one hundred miles south of where we made our discovery. Buratovich was sent to Patagonia to conquer land inhabited by the native peoples, and like other regrettable military campaigns in other countries, this one resulted in the systematic extermination of most of the native tribes in Patagonia.

The bones discovered by Buratovich included tailbones and ribs, which were identified as belonging to dinosaurs by Florentino Ameghino, a prominent Argentine paleontologist and internationally renowned scientist of his time. He is widely recognized as the "Father of Vertebrate Paleontology" in Argentina. News and publication of the discovery in 1883 set off a wave of further collecting. Florentino sent his younger brother Carlos, an expert collector seasoned by the rough winters of southern Patagonia, to find more. At about the same time, a newly created museum in La Plata, a town about forty miles south of Buenos Aires, hired another famous fossil collector named Santiago Roth. Roth was a Swiss scientist interested in dinosaurs and fossil mammals, as well as the geology of Patagonia. Over three decades, Roth and other scientists made important collections of dinosaurs and other fossils that comprised the ancient fauna around Neuquén.

Over the next century, dinosaur fossils were discovered in several parts of Argentina. These fossils dated from throughout the Mesozoic era—an interval of geologic time often called the Age of Large Dinosaurs. The Mesozoic is divided into three periods. The Triassic period extended from about 250 million years ago to about 206 million years ago. The Jurassic period began at the end of the Triassic and concluded about 144 million years ago. The final period, called the Cretaceous, extended from 144 million years ago to 65 million years ago.

In 1884, bones that Roth had discovered were sent to Richard Lydekker, a famous British paleontologist. The scientific study of dinosaurs was in its incipient stages, even though their fossils had been collected for centuries and had probably inspired the creation of numerous mythological creatures, such as the griffin. In scientific terms, dinosaurs had first been recognized as ancient reptiles, in England, during the 1820s. By the 1880s, few skeletons of dinosaurs had been discovered, collected, and studied. Nonetheless, Lydekker identified the remains Roth had found as belonging to titanosaurs, long-necked, giant, plant-eating dinosaurs belonging to the group called sauropods. Lydekker noticed the remarkable similarity between these Patagonian dinosaur bones and others found in 70-million-year-old rocks from India, which belonged to a titanosaur that he had named *Titanosaurus indicus*, which means "Indian *Titanosaurus*." He thought that the Patagonian species and the Indian species were closely related, and thus he christened the Patagonian species with a new name, *Titanosaurus australis*, which means "southern *Titanosaurus*." Later studies have shown that the bones identified by Lydekker were not so closely related to the Indian titanosaur, but instead belonged to a quite different and older titanosaur from the 80-million-year-old, Cretaceous rock unit called the Río Colorado Formation. This is the same rock unit that produced the fossils that we discovered more than a century later.

GEOLOGIC AND
EVOLUTIONARY TIME SCALE

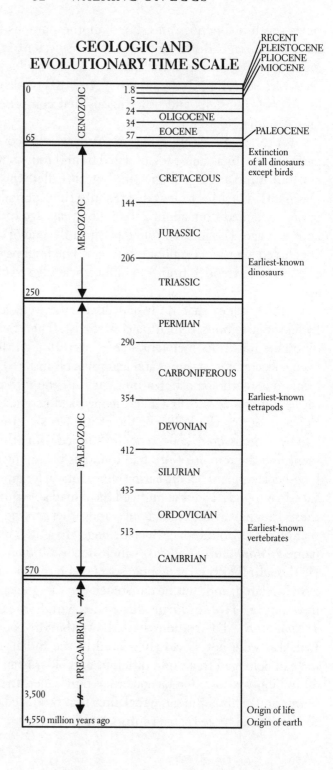

Lydekker was also the first paleontologist to recognize the presence of meat-eating theropod dinosaurs in Patagonia, based on fragmentary remains of fossil skeletons. At the turn of the century, another British paleontologist, Arthur Smith Woodward, studied a more complete, yet still fragmentary, skeleton of a theropod from Patagonia. Its remains included portions of its imposing jaws, which were found by Santiago Roth in rocks of the Cretaceous period from the province of Chubut. Although Woodward named it *Genyodectes serus*, its fragmentary nature prohibits us from identifying it more precisely than to say that it is one of the large meat-eating theropods.

Among the early discoveries made in the 1900s was a fragmentary lower jaw of a horned dinosaur, related to *Triceratops* and its kin, from late Cretaceous rocks of Chubut. To date, this remains the only bone of a horned dinosaur ever found in South America. Although its fragmentary condition raised some doubts about whether it really belonged to a horned dinosaur, the discovery of footprints from horned dinosaurs in Bolivia has provided some support for this identification. However, the shape of a footprint can vary tremendously depending on whether the animal walked across wet or dry sand or mud, making them less reliable when identifying extinct animals.

Santiago Roth and other collectors from the La Plata Museum amassed one of the largest early collections of dinosaurs. Yet, the importance of Argentina's dinosaur fossils was not internationally recognized until the 1920s, when a German paleontologist named Friederich von Huene—one of the most prominent fossil experts of the early twentieth century—published several detailed studies. He documented that a wide range of dinosaurs had once inhabited Patagonia, including meat-eating theropods, giant sauropods, and horned dinosaurs. But curiously, instead of catalyzing further collecting and research, the seminal work of von Huene arrested scientific investigations of dinosaurs in Argentina until the late 1950s.

At that point, an Argentine biologist interested in evolution, Osvaldo Reig, initiated a multiyear collecting program in the inhospitable landscape formed by Triassic rocks in northwest Argentina. Among the numerous vertebrate fossils discovered and studied during Reig's expeditions were those of the primitive meat-eating dinosaur *Herrerasaurus*, one of the oldest dinosaurs known. Reig described the skeleton of this menacing, ten-foot-long carnivore in 1963.

José Bonaparte, a prolific Argentine paleontologist who is still discovering and studying new dinosaurs in the country, continued Reig's work. Bonaparte and his crews, which often included Luis, explored the extensive dinosaur-bearing rock layers of Patagonia. These crews discovered the first Triassic dinosaur eggs and hatchlings in Santa Cruz; the first well-known South American Jurassic dinosaur fauna in Chubut; and a myriad of Cretaceous dinosaurs in Río Negro and Neuquén.

The systematic exploration of Patagonia's rich dinosaur deposits, initiated by Bonaparte, rests today in the hands of his current and former students, such as Luis, who through Argentine and international expeditions are unraveling the secrets of dinosaurian history in Patagonia. In fact, vertebrate paleontology has become one of the most internationally recognized scientific disciplines conducted in Argentina. Especially important for our expedition was the recognition of a new group of large, meat-eating dinosaurs called abelisaurs by Bonaparte and his former student Fernando Novas. The most complete and spectacular skeleton belonged to *Carnotaurus*, a thirty-foot-long biped that featured a tall skull with prominent horns over its eyes. However, its arms were quite short in relation to the rest of its body, raising questions about what if anything the animal had used them for.

All of this work set the stage for our expedition because the discoveries of large dinosaurs made by previous workers would play an important role in our own investigations. But we were actually in pursuit of much smaller quarry.

CHAPTER TWO

What Were We Searching For and How Did We Decide Where to Look?

The most common components of dinosaur research are the discovery of previously unknown species and the analysis of their diversity and evolutionary history. Other important biological components are the study of how dinosaurs changed through time, as well as how they moved and behaved. These intriguing scientific investigations, however, must be founded on a clear understanding of the genealogical relationships of each species. Without knowing the origin of each group, which represents the starting point for later physical transformations, it is impossible to reconstruct how a new anatomical structure or behavior came to be. Related geological research involves estimating the age of the rocks that contain dinosaur fossils and determining what the rocks can tell us about the climates and environments in which the dinosaurs lived.

Fossilization is a natural geologic phenomenon, and fossils are found worldwide. But to discover new dinosaur species, a paleontologist can't just fly to some far-off part of the globe and start digging, because the chances of finding previously undiscovered species are quite low. In fact, paleontologists usually find new species in a previously unexplored place where rocks of the right age are exposed. Consequently, dinosaur expeditions are preceded by a careful examination of several possible areas for exploration, with the final decision based on the goals of an underlying research program as well as an assessment of the accessibility and risks of the selected areas.

Every year, international expeditions scout some of the most inhospitable regions of the world for dinosaur bones, from the deserts of northern Africa and central Asia to the frozen peaks of Antarctica and Alaska. Numerous smaller expeditions are conducted in almost every country. In our case, we knew that Patagonia had produced an incredible assortment of dinosaur fossils over the previous century. The scientific reports describing those dinosaurs also contained information about where they were found, the kinds of rocks entombing them, and their approximate age. In addition, we knew where our colleagues were conducting explorations and finding new kinds of dinosaurs throughout the region.

Yet in spite of all the earlier and contemporary fossil discoveries, most parts of Patagonia are still virtually unexplored by paleontologists, partly because of the region's size and partly because of its desolate landscapes. Most modern inhabitants of Patagonia live in small towns scattered over large stretches of sparsely populated desert. Aside from the highways that connect these widely separated towns, roads are rare and unpaved. Oil companies built most of these dirt roads to facilitate their search for oil and natural gas. In the wake of these geological prospectors followed smaller field crews of paleontologists prospecting for fossils. But fossil exploration in Patagonia today is similar to paleontologic exploration in the American West at the turn of the century, when famous dinosaurs such as *Tyrannosaurus* and *Diplodocus* were first discovered.

One of these unexplored regions is near the center of the province of Neuquén, some six hundred miles southwest of Buenos Aires as the crow flies. There, vast badlands formed from rock layers deposited near the end of the Mesozoic era are exposed around an extinct volcano called Auca Mahuida. We knew that superb fossils of ancient birds and their closely related dinosaurian cousins had been found in comparable rock layers not far from this area because Luis had studied them while compiling his dissertation on ancient Mesozoic birds from South America. Over the last few years, the origin and early evolution of birds has become one of the most active areas of research in vertebrate paleontology. Yet, except at a few unusual sites, remains of these ancient birds are extremely rare, and as such, they represent highly prized discoveries for paleontologists.

So, in 1996, Luis and some members of our field crew undertook a brief exploration of the rock layers exposed around Auca Mahuida at the end of a field season to central Argentina. The goal of our initial expedition was not to find skeletons or eggs of large dinosaurs, but rather to search for the delicate fossil bones of the birds and their dinosaurian relatives that inhabited Patagonia near the end of the Mesozoic. Instead, Luis and his crew discovered some fossil bones of giant, plant-eating sauropods, as well as fragmentary remains of other dinosaurs and creatures that had lived with them. Even though no fossil birds were found, the rocks of this region looked promising and deserved more of our attention. This short foray into the region around Auca Mahuida proved crucial to our later success in that we gathered important information about the rocks of the area and identified where we could find water, food, and shelter. We also began to familiarize ourselves with the roads and tracks that might allow us to gain access to promising outcrops.

After returning from the 1996 trip, we decided to organize another expedition in 1997. This field trip would include a more extensive, monthlong exploration of the Auca Mahuida area with a larger crew. Our initial goal was the same as that of the previous year: to find the remains of ancient birds and the small, meat-eating theropod dinosaurs that are closely related to them. But even if we didn't find any birds, we knew that a more complete collection of the ancient animals from this region, as well as a thorough study of its rocks, could enhance our understanding of the age of these birds and the ecosystem they had inhabited.

As with all expeditions, numerous challenges faced us before we could embark. We had to draw up detailed plans and make agreements with the governments and paleontologic institutions in the areas where we hoped to work. We also needed to find funding for tents and other camping gear, cooking equipment, food, drinks, collecting tools and materials, renting and maintaining vehicles, and airline tickets.

Suzi Zetkus and Luis took charge of flight arrangements, buying supplies, and renting our vehicles. This was an enormous job. Suzi volunteers countless hours of her leisure time to tour visitors at the American Museum of Natural History. In addition, she has participated in several paleontological expeditions both for the museum in New York and for other institutions. She served expertly as our expedition coordinator as well as an indefatigable driver, keeping track of supplies and running errands between our camp and Neuquén, the nearest large city. Detailed lists of the supplies we would need had to be compiled and each item obtained. Some items were purchased in Argentina, but others had to be bought in New York and shipped to Buenos Aires on our flights.

Our camping and cooking equipment was fairly standard. But for celebratory *asados*, a special kind of barbecue commonly practiced by the Argentines, we packed a grill and a couple of iron rods to build a cross on which to hang the carcass of a sheep or goat over the fire.

Our list of collecting equipment was more specialized. To excavate fossils and rock samples, we shipped shovels, picks, rock hammers, crowbars, chisels, brushes, special glue, dental tools, and sample bags. To construct protective jackets around fossils, we bought large bags of plaster, plaster bandages, burlap, and dozens of rolls of toilet tissue. We had to accurately record the location of our fossil sites and places where we collected rock samples, so we ordered geologic maps and GPS (global positioning system) units. These small computers receive signals from orbiting satellites for calculating the precise longitude and latitude of the place where one is standing. A special instrument called a Brunton compass was required to measure the thickness of the rock layers that formed the ridges from which we collected samples for magnetic analysis. These samples had to be wrapped in aluminum foil and secured with masking tape for the trip back to the lab.

In developing the collecting agreements with the local governments and institutions in Neuquén, we received invaluable assistance from our colleague and co-leader Rodolfo Coria. Luis and Rodolfo had become friends when Rodolfo worked as an illustrator and collector for José Bonaparte at the same time that Luis was Bonaparte's student. Rodolfo is a tall, lanky, easygoing man who now serves as the director of the Carmen Funes Museum, located in the town of Plaza Huincul, about an hour's drive west of Neuquén. He had moved to Plaza Huincul from Buenos Aires to be closer to the rich dinosaur deposits of Patagonia. Rodolfo and his team of fossil hunters from the Carmen Funes Museum have helped to discover and describe many new dinosaurs from Neuquén, including one of the largest animals ever to walk on earth, *Argentinosaurus*, and a ferocious meat-eating dinosaur,

Giganotosaurus, which rivaled *Tyrannosaurus rex* in size. Through Rodolfo's tireless efforts, the Carmen Funes Museum now occupies a prominent position on the international map for vertebrate paleontology.

In the early decades of fossil collecting, permits were not required, and foreign crews simply went into other countries and took title to whatever they found. Today, however, most countries have understandably passed laws to protect their paleontological heritage from foreign exploitation. Thus, before one travels to another country to collect, one must get permission in written agreements or permits that stipulate how the fieldwork and research will be conducted. Our agreement with Rodolfo and the Argentine authorities allowed us to prospect and collect in all the areas that we were interested in. If we were fortunate enough to find fossils, the agreement allowed us to bring the fossils to the United States to clean, cast, and study them. After the research for scientific articles was completed, the specimens would be returned to Argentina and housed at the Carmen Funes Museum. Such arrangements are now commonly struck between museums and universities in different countries.

To fund the expedition, Luis wrote a grant proposal to the National Geographic Society, describing how we wanted to prospect for fossils around Auca Mahuida. Such proposals must document the goals of the project and the potential for finding new specimens that could improve our scientific knowledge. Based on the exploratory trip taken in 1996 and the wealth of new material being discovered in the region, we believed that a grant to explore the area in more detail could be justified. As part of the proposal, Luis had to itemize a budget for our equipment and travel, and after reviewing the proposal, the Society approved it. In addition, Lowell raised money from the InfoQuest Foundation in California. In all, we raised almost $20,000 for the expedition. Rodolfo and his colleagues from the Carmen Funes Museum provided additional vehicles and equipment.

After many months of homework and planning, we were at last ready to search for ancient fossil birds and other animals in the remote badlands of Patagonia.

CHAPTER THREE

The Scene
of an Ancient Catastrophe

Auca Mahuevo

During the last week of October in 1997, Suzi Zetkus and Luis flew from New York to Buenos Aires to buy the rest of the supplies we needed and to rent a van for transporting the gear and crew into the field. Lowell and the rest of the crew flew down during the first week in November, and we met several students and scientists from universities and museums in Argentina who would also be part of our expedition. After completing the shopping for the last items on our list of supplies, we were ready to begin the seven-hundred-mile drive to Auca Mahuida.

Our expedition left Buenos Aires on the afternoon of November 6. We traveled in two vehicles, an aging pickup truck that belonged to Luis's sister, which carried most of the field gear, and a more luxurious and air-conditioned van, which carried most of the crew members. For hours, we drove through a region of flat plains usually called the Pampas, the agricultural heartland of Argentina. We stopped at about nine o'clock in the evening when we arrived at the small rural town of Pehuajó, about one-third of the way to Auca Mahuida. Marked by a colossal statue of a standing tortoise, borrowed from a famous Argentine song narrating the adventures of the audacious *Manuelita*, Pehuajó enjoys the laid-back spirit of its reptilian denizen. The vehicles were running well, which is always a relief, and we were pleased to have gotten off to a good start. After checking into a small hotel for the night, we ate dinner at a local restaurant and went to sleep. We would not enjoy the luxury of a bed again for about four weeks.

The next day we spent driving, hoping to make it all the way to Auca Mahuida by nightfall. Despite the long hours in the vehicle, the sights along the way, especially the birds, were entrancing. El Niño had brought an unusual amount of rain to the Pampas, which had formed shallow ponds and lakes along the side of the road that attracted many waterbirds. Identifying the flamingos, spoonbills, coots, ducks, and various raptors kept us entertained as the miles rolled by. As we drove farther west, the land grew drier and more rugged. Near Santa Rosa, about halfway to Auca Mahuida, we drove along enormous ridges that represented ancient sand dunes deposited by huge sandstorms near the end of the last Ice Age, ten thousand to twenty thousand years ago. However, we were on a quest to find older rocks, ones laid down as South America had split apart from Africa more than 70 million years ago, when large dinosaurs ruled the continent.

As we stopped for gas in Santa Rosa, we were reminded of just how small a world we live in. Buying gas at the pump next to ours was François Vuilleumier, a curator in the Ornithology Department at the American Museum of Natural History, the same department in which Luis worked. Nattily dressed in duds that would be the envy of any gaucho, he had been collecting data on the birds of the region for his research. In fact, a third member of the Ornithology Department was present. Paul Sweet, who has traveled all over the world to acquire new specimens for the museum's collection of birds, was accompanying our expedition as a scientific assistant. Even more bizarre, François had just driven Sara Bertelli, an ornithology graduate student from the University of Tucumán in northwest Argentina, who is also a student of Luis's, to Rodolfo's museum so that she could join our expedition to collect specimens of a primitive bird called the tinamou. Ironically, scientists go to the field to get away from the responsibilities at the museum, but as chance would have it, Luis, Paul, and François constituted a quorum for an impromptu departmental meeting at a gas station halfway around the world.

Finally, in the late afternoon, we met Rodolfo Coria and his crew members at a small town in northwestern Patagonia called Barda del Medio. From there, we drove the last seventy miles to our field area as the sun set over the extinct volcano at Auca Mahuida. As darkness descended, we set up our camp at a *puesto*, a small ranch that serves as the home for rugged livestock ranchers who raise sheep, goats, cattle, and horses. This particular *puesto* was owned and managed by Doña Dora, a tough but friendly seventy-year-old native of this region, and her partner Don José. Along with Doña Dora's grandson Josecito and a handsome young gaucho named Juan, they tended herds numbering several hundred animals. Their generous hospitality and good company would prove essential to the success of our expedition.

Occasionally, we would buy one of their goats or sheep to cook slowly over a bed of coals as part of an Argentine *asado*. These were special occasions that often turned into raucous parties with Doña Dora and her family. While we ate our dinner that night, stories abounded about other places we had visited to collect fossils and events on their *puesto*.

Life on the *puesto* is difficult. Resources are scarce, and each member of the *puesto* family, including the dogs that help tend the livestock, bears heavy responsibilities. The dogs were clearly not pets, and our crew had to be careful not to become emotionally attached to them. One had an adorable litter of puppies soon after we arrived. Later, when we returned from a day of prospecting for fossils, we discovered that Doña Dora had drowned them. Her motive, as she explained, was simple. There wasn't enough food to feed them. Dangers lurked among the ridges and ravines of the ranch. We were amazed by a story Doña Dora told about clubbing a puma to death in self-defense. Luis encountered his own brush with disaster one day as he was returning to camp. The gauchos had just finished castrating a bull as Luis approached, and not seeing Luis, they turned the bull loose and ran for cover. The bull, noticeably irritated at the treatment he had received, arose to see Luis ambling peacefully toward him. The bull then charged, and Luis's only means of escape was to sprint for a hedge of thornbushes that formed the fence of the corral. He leapt into the wall of thorns as far as he could. With the bull snorting at his heels, it was a narrow and painful escape, but far preferable to the alternative.

Our dinners with Doña Dora's family typically lasted late into the evening, and the evenings were spectacular, with crimson sunsets followed by starlit skies accented by an occasional streaking meteor or passing satellite. There was no Big Dipper with its North Star because they are not visible from the Southern Hemisphere. Instead, the picturesque Southern Cross kept watch over us.

Many of the crew born in the Northern Hemisphere had never seen the southern constellations before and were looking forward to the opportunity. One had brought a star chart to help identify them, but after a few seconds of examination, Lowell realized that it was a chart for the Northern Hemisphere, which was useless here. Lowell concluded that he and his colleagues from the United States didn't actually live in the Northern Hemisphere, but rather in the "moronosphere." Fortunately, the native Argentines were still willing to serve as our mentors for stargazing.

Most of these starlit nights were cool but not cold—good sleeping weather. Although it doesn't rain very often during the time of the year we were in Patagonia, the wind blows almost constantly. Fifty-mile-an-hour gusts, capable of collapsing a tent or sending it careering across the rocky ridges and ravines, are not uncommon. Occasionally, local inhabitants, such as tarantulas, would pass through if you didn't keep your tent zipped up, not to mention the geese, chickens, baby goats, and dogs that literally owned our backyard camp.

The *puesto* had no running water, so there were no toilets or showers. The bathroom was behind some distant bush or rocky outcrop, and we washed up in the small stream that flowed in the riverbed below the *puesto*. Although the stream was only a couple of inches deep, we dug a hole big enough for washing the dust off at the end of the day, but we did not drink the water from it. A recent flood had destroyed Doña Dora's well, and the livestock had contaminated the stream's water. So every few days, a couple of crew members drove back into Neuquén to buy groceries and fill our water containers.

Cities like Neuquén, which with its suburbs has a population of several hundred thousand people, have all the conveniences of a modern city, including gas stations, Laundromats, and a supermarket that covers an entire city block. We took advantage of all these facilities—especially the shower room at one of the local hotels—and although we did not have all these comforts of home at Auca Mahuida, we were quite content and ready to begin our work.

After a good night's sleep, we awoke with great anticipation. At long last, we could begin prospecting for fossils in the ancient layers of rock that formed the gorgeous cliffs behind the *puesto*. Looking for fossils usually involves a lot more walking than it does digging. There are two basic kinds of fossil collecting, prospecting and quarrying. When one first begins searching for fossils in a new area, as we were, one starts by prospecting. As the name implies, this involves walking over promising-looking ridges, flats, and ravines while looking for small fragments of fossil bone that are weathering out of the rock layers on the surface of the ground. These fragments are clues that a dinosaur skeleton may be buried underneath the weathered surface of the rock. A crucial skill is to be able to distinguish between the ancient rock layers and younger layers or debris exposed on the surface. Although the ancient rocks that entomb dinosaur skeletons are often exposed in bare ridges and ravines, sometimes they are partially covered by much younger soils. So, inexperienced collectors frequently spend hours searching in these younger soils only to find the bones of a cow that died last winter—a very discouraging experience.

The color and mineral content of fossils varies from place to place, depending on the chemical reactions that occurred as the bones and teeth became petrified. When collecting in a new area, a good practice is for the first person who finds a fossil fragment to pass it around so that other collectors can begin to develop an image of what to look for. That's why we assembled the crew to examine the dinosaur eggs immediately after we first discovered them. Finding fossils requires patience and determination, and sites may yield their fossils only after having been examined two or three times.

After finding fragments of fossil bone weathering on the surface, it is critical to look for their source. Sometimes, one can follow a trail of small fragments that have washed down a hillside right up to where an entire dinosaur skeleton is buried beneath the surface. It's like a detective following drops of blood to locate a body, and this approach would lead us to some of our best finds. In other instances, whole days can be spent prospecting for fossils without ever finding one splint of bone, let alone a complete skeleton in the ground. While we prospected this first day, we were serenaded by the screeches and squawks of cliff-dwelling parrots that lived along the rocky ridges around the *puesto*.

If one is lucky enough to find fragments that lead to a good fossil, it must be excavated and transported to a museum for final cleaning and study. Excavating a fossil is called quarrying and means digging around a bone or skeleton that is buried near the surface of the ground and encasing it in a protective jacket of toilet tissue and plaster bandages. First, tissue is placed over the fossil bones and moistened to form a protective layer between the fossils and the plaster bandages, and then the bandages are applied. Once the bandages dry, the jacket can be lifted out of the ground and transported safely back to a museum without damaging the fossils inside. The time required to quarry out a dinosaur skeleton often depends on the hardness of the surrounding rock and the size of the specimen. During this expedition we would eventually need to both prospect and quarry.

Our first day in the field we uncovered nothing new and exciting. Luis and his crew found scraps of fossil bone: some fragments of turtle shell, small pieces of shell from a dinosaur egg, and armor plates from ancient crocodiles. Rodolfo and his team found the tail section from the skeleton of a small sauropod called a titanosaur, not a bad discovery but not significant enough for a team of collectors to spend a whole week excavating.

By looking at the clues in the rock layers, Lowell noticed a good geologic reason for why we were finding only fragments of bone. The layers of rock around the *puesto* were composed primarily of cemented sand, gravel, and even small boulders that had been deposited by fast-running rivers or streams on an ancient alluvial fan or wedge of debris that lay fairly close to some small mountains or hills. Alluvial fans are found at the mouth of many mountain canyons today. Lowell knew that swift, turbulent currents are required to carry such large objects and that these would destroy most skeletons of dinosaurs and other animals being carried along in the river. So, these rock layers, although beautiful, were probably not the best place to look for well-preserved fossil skeletons. Nonetheless, these layers of sand and gravel sometimes contain pockets of mudstone and siltstone, deposited by slow eddies in quiet parts of the channel, where more complete fossil skeletons can be preserved. However, from the top of outcrops near the *puesto*, we could see another extensive set of glowing reddish brown badlands off in the distance, and we thought that they might contain more complete fossil skeletons, if we could find a way to get to them.

On the morning of November 9, we drove off in that direction. As we bounced back toward the main road from Neuquén, we passed a gap in a small ridge, and through the gap, we caught a glimpse of the badlands we were searching for. Fortunately, a small dirt road ran through the gap, so we rumbled a mile or two down into the center of the sunlit layers of ancient rock.

We parked the vehicle by the side of the road and turned the crew loose to prospect the area for an hour or two. Most forays like this are unsuccessful because few, if any, fossils are found, so we had no great expectations. As we walked out onto the flats adjacent to the beautifully banded layers of sandstone and mudstone, we scanned the ground, searching for scraps of fossil bone. Within five minutes of leaving the vehicles, we found ourselves walking across vast fields littered with dinosaur eggs. Acres and acres of reddish brown mudstone were exposed on the flats, and every few steps, a cluster of broken eggs lay perched on the surface. There were eggs everywhere, tens of thousands of them. Although these had crumbled to pieces, the complete eggs were clearly about a half foot in diameter, and the eggshell was about one-tenth of an inch thick.

The first question, of course, was what kind of dinosaur had laid those eggs? Paleontologists had often assumed that large eggs such as these had belonged to the colossal sauropods. But we had worked in the Gobi Desert of Mongolia with a crew that had discovered a meat-eating *Oviraptor* embryo inside an egg that had previously been thought to belong to a plant-eating *Protoceratops*. That discovery had changed seventy years of accepted paleontologic wisdom and made us cautious about prematurely identifying the kind of dinosaur that had laid the eggs at our site. We did not want to misidentify the eggs and create the kind of confusion that had surrounded the eggs from the Gobi for so long.

We knew that sauropods had lived in this region of South America near the end of the Mesozoic; we had found parts of their skeletons just the day before in the outcrops of cemented sandstone and gravel near Doña Dora's *puesto*. For more than a century paleontologists have recognized that sauropod dinosaurs represent the largest land animals ever known. Their names roll off the tongues of children and adults alike, *Apatosaurus* (formerly *Brontosaurus*), *Diplodocus*, *Brachiosaurus*, and so on. Adult remains of these animals have been collected on numerous continents, but no skeletons of embryos inside their eggs had ever been found. In fact, some paleontologists had speculated that these gigantic dinosaurs had given birth to live young rather than having offspring that hatched from eggs. The ultimate clue needed to identify what dinosaur had laid these Patagonian eggs was still missing. We gathered excitedly to assess our initial discovery, and Luis issued his challenge for someone to find an embryo before we once again fanned out across the flats and ridges.

Fossils of embryos rank among the rarest of dinosaur remains because they represent the fragile skeletons of baby dinosaurs that have not yet hatched out of the egg. The skeleton of a growing embryo is only partially made of bone. Much of it is still composed of softer cartilage. This cartilage is rarely fossilized because it often decays along with the skin, muscles, and organs soon after the embryo dies. Consequently, such small, delicate skeletons are hardly ever preserved. They either decay or get destroyed before they become fossilized. Yet, this huge treasure of fossil eggs was also encouraging because we suspected that a few of them could easily have embryonic bones preserved inside.

A half hour later, Carl Mehling approached us excitedly. Carl was one of our designated fossil collectors from the American Museum of Natural History. A passionate collector, his enthusiasm and acute sense of humor were infectious, and he often instigated hilarious dialogues that made him the life of the party, but at this moment his face reflected a measure of ecstatic anticipation. He had found an egg with a small, rocky sheet of bumpy, mineralized material preserved inside, and he thought the texture preserved on the surface resembled dinosaur skin. At first, Luis was skeptical. The patch of textured rock was small and might just have been some unusual mineral crystals. Besides, fossils of dinosaur skin and other types of soft tissue are extremely rare, and no one had ever discovered fossils of embryonic dinosaur skin inside an egg. To make a positive identification, we would have to keep searching for more.

By the end of the day, we were exhausted but elated. We knew we had discovered a remarkable new fossil site, even though we had not found any more patches of possible skin and were not sure exactly what kind of dinosaur eggs we had found. Now, it was time to celebrate.

Next morning when we returned to the site, most of the crew continued looking for more fossils of the potential embryonic skin and some embryonic bones. Luis and a few others tried to estimate how many eggs and nests were exposed on a small portion of the flats by tying colored tape to the branches of some small bushes to mark out a trapezoid and measuring the distances between bushes using the GPS. With this information, they calculated that the two sides were about 1,000 feet long, with the top about 250 feet across and the base about 400 feet. Within that area they counted about 195 clusters of eggs. Later, we would refine the estimates, but the number of egg clusters would still be extraordinary.

The abundance of eggs triggered a lighthearted discussion among our crew members regarding what we should name our new fossil site. After brainstorming over several humorous possibilities, we all agreed on a name—Auca Mahuevo. In part, the name represented a pun on Auca Mahuida, but it also acknowledged the seemingly countless number of eggs preserved at the site. *Mahuevo* is kind of a Spanish contraction for *más huevos*, which means "more eggs."

As part of our prospecting for eggs and embryos, we walked many miles over the flats and adjacent ridges. The layer with eggs seemed to go on forever. Many thousands of nests were spread over several square miles. But all the eggs seemed to be restricted to a single layer of rock.

Noting this relationship, Lowell began studying the rocks that contained the fossils, intent on finding clues that would shed more light on the dinosaurs that had lived at Auca Mahuevo. The beautiful reddish brown mudstones, as well as distinctive greenish sand layers that were mixed in with them, might provide important evidence to help us interpret what kind of environment the dinosaurs were living in, as well as how long ago they had lived.

Lowell's most important geological job was to start at the bottom of the sequence of rock layers and measure the thickness of each layer of sandstone and mudstone. As the measurements were taken, he drew a picture of the different rock layers in his field notebook. That depiction is called a stratigraphic section. These drawings are important for telling time back when the dinosaurs were living at the site. Based on previous work done by other scientists, Lowell knew that the rocks at Auca Mahuevo were deposited sometime during the latter stages of the Cretaceous period, between about 70 million and 90 million years ago. But his challenge was to find evidence that would specify the age more precisely within that long interval.

Meanwhile, our crew of collectors was having more success finding fossils. Two eggs of great importance were discovered, containing fragments of embryonic bones. Our ace collector of these kinds of fossils turned out to be a talented young woman, Natalia Klaiselburd, an undergraduate biology student with an interest in paleontology. She possessed the keen eyes of an eagle and excelled at finding egg fragments that contained microscopic fossils inside them. The fragments of bones found in the two eggs mentioned above were still not large enough to allow us to identify what kind of dinosaur had laid the eggs, but they indicated that if we kept looking, we might find eggs that did. We were already suspecting that the place could go "big time."

The next day was filled with exhausting work. We checked thousands of fossilized egg fragments strewn across the surface of the flats and in adjacent ravines, searching for embryonic bones. Although the vast majority contained no embryonic remains, Luis found one that contained a large patch of mineralized bumps, like the small patch Carl had found during our first day at the site. The surface of this patch was ornamented with scaly-looking bumps crossed by a triple row of larger and more-rectangular-shaped plates. This egg left no doubt: we had discovered the first fossils of embryonic dinosaur skin ever found. Members of the crew hugged each other, sharing the thrill of exhilarating success in a moment of tremendous excitement and elation, one that Luis would remember his whole life.

We were overjoyed with our good fortune, but it still took a while for the significance of our discovery to sink in. Little is known about dinosaur embryos, and with our discovery, we had found one of the most important missing pieces of the scientific puzzle about these long-extinct animals—what they looked like when they were first born. Furthermore, the skin on the embryos that our crew had found was not preserved as an impression left by the skin in the surrounding mud, as is usually the case. Our fossils represented three-dimensional replicas of the actual skin of the embryo, the only direct evidence we have of what dinosaurs looked like on the outside.

We had been in the field only four days, yet our expedition had already been successful beyond our wildest dreams. As we drove back to camp at the end of the day, we reflected on all the work that lay ahead, especially the need to find more embryonic bones that would help us identify for sure what kind of dinosaur had laid the eggs. But that could serve as the focus for tomorrow's work; tonight we would celebrate our success with some fine Argentine wine, a tasty *asado* featuring beef and goat, and a lively dance party.

We resumed our search for embryos the next morning, but with a slightly different strategy. Up to this point, we had been concentrating primarily on the eggs exposed on the flats adjacent to the ridges in the badlands, but now we decided to expand our search to the badland ridges and ravines themselves. About a half mile from the flats, a couple of Rodolfo's crew members, Pablo Puerto and Sergio Saldivia, found a hillside, below a ridge, littered with eggshell fragments. Pablo is the chief preparator at the Egidio Feruglio Museum in the province of Chubut. His vivacious personality and quick sense of humor greatly enlivened the atmosphere of our camp, and his skills as a collector proved essential in helping us identify the victims of the ancient catastrophe at Auca Mahuevo. Sergio Saldivia is Rodolfo's chief preparator at the Carmen Funes Museum. Quiet by nature, his talent as an *asador* brought great satisfaction to our hungry field crew, and his skill as a preparator proved essential in cleaning the eggs and bones of the dinosaurs so that we could study and identify them.

Quarrying in under the surface of the reddish brown mudstone, Pablo and Sergio discovered several complete, well-preserved eggs. When they began to probe into one of the eggs, some small, thin, brown bones appeared. Our quest to find an embryo had finally been successful. The bones, rather large for embryonic bone at about three to four inches long, appeared to fit up against one another, and their shape suggested that they formed the leg of a baby dinosaur. Although we could not unequivocally identify them in the field, we were pretty sure we would be able to say which dinosaur they had come from when we got back to the museum laboratory and prepared them properly.

Lowell began to investigate where the new embryo quarry fit in the sequence of rock layers exposed along the ridges and across the flats. Assisting in this enterprise were Julia Clarke and Javier Guevara. Julia is a graduate student studying under Jacques Gauthier in the Department of Geology and Geophysics at Yale. Jacques is a close friend and colleague whose research has greatly improved our understanding of the evolution of meat-eating dinosaurs and their evolutionary links with birds. Julia actually played a dual role in the activities of the expedition. In addition to collecting fossils, her geological background made her a valuable asset in helping us to unravel the environmental mysteries at the site and refine the time period during which the dinosaurs lived. Javier Guevara is an undergraduate geology student. In addition to helping us collect fossils, he also assisted in collecting and documenting the geologic data that would paint a picture of the ancient environment that the dinosaurs lived in.

Lowell's geological team faced a basic question. Was the embryo quarry located in the same rock layer that was producing eggs a half mile away at the flats, or was it contained in a different egg-producing layer? If it was in the same layer, all the eggs would have been laid at about the same time, but if it was in a different layer, the site would contain at least two nesting grounds that had been inhabited at different times. To find the clues to solve this mystery, we had to walk on the egg-producing layer all the way from the quarry back to the flats. We traced the layer that contained the eggs across the rugged ridges and ravines of the badlands back to the area around the flats where Lowell had measured the stratigraphic section. It was easy to follow the layer because of its dark reddish brown color, muddy texture, and the eggshell fragments that had weathered out on its surface. After a half hour of careful hiking, we established that the quarry was in the same layer of mudstone that had produced fossilized eggs on the flats. So it seemed as if Auca Mahuevo represented just one enormous nesting site.

The wind blew mercilessly throughout the night of November 12 and all through the next day. Patagonia was testing our mettle with a blast of nasty weather. Gusts as high as fifty miles per hour stormed across the dusty landscape, sandblasting everything in their path, including our eyes. Many of our tents were blown over or damaged, so we had to spend much of the day tending to camp.

When the weather calmed, we returned to our work. Over the next two weeks, we collected dozens more eggs, including both fragments with patches of fossilized skin from the flats and clusters containing bones of embryos from the quarry in the badlands. In several instances, collecting the clusters of eggs required the preparation of large plaster jackets to protect the eggs during the trip back to the museum. The primary responsibility for this operation fell to Pablo, Sergio, and Marilyn Fox. Marilyn is an excellent fossil preparator at Yale University for Jacques Gauthier. We had known Marilyn for many years because she had worked with us at the American Museum of Natural History in New York. On this expedition, she was our chief fossil excavator, and her immense talents as a preparator would prove critical because later she assumed the responsibility for preparing the delicate fossils preserved inside some of the eggs.

Some of the blocks of mudstone we collected contained more than twenty eggs. Before transporting them, Luis tried to put our minds at ease by assuring us that the blocks were not much larger or heavier than a typical *jamón*—a Spanish ham. With the protective plaster bandages that enveloped them, however, some actually assumed the proportions of a healthy sow, weighing in at several hundred pounds. As you might imagine, these turned out to be difficult to move, even with a lot of people helping to pull and lift. Furthermore, as is often the case in the field, we could not drive our truck all the way to the place where we had found the eggs. To get the heavy blocks down the hill to the truck, Luis and Rodolfo borrowed a large sheet of scrap metal from Doña Dora. After punching some holes at each corner of the sheet and attaching ropes, they had a makeshift "sled" on which we could put the blocks and slide them down the hill. In deference to Luis's original description of the blocks, this contraption was affectionately dubbed "the ham luge." To move the blocks, some crew members pulled on the ropes in front, while others steadied the block and the sled with the ropes attached to the back. It took almost all our crew members to move the blocks about fifty yards from the quarry down the hill to the trucks. Lifting them into the back of the pickup also proved challenging, but with everyone's help, we managed it.

Following the move, Lowell collected more rock samples in hopes that they might contain fossilized microscopic pollen. The shape of a pollen grain is quite specific for each species of plant, and particular assemblages of fossil pollen are often restricted to narrow intervals of geologic time. If we were fortunate enough to find fossil pollen in the rocks at the site, this would provide clues about the kinds of plants that lived on or near the floodplain where the dinosaurs had laid their eggs, as well as about the age of the eggs. However, such analyses would once again have to wait until we got the samples back to the laboratory.

We spent a few more days looking at exposures of rocks in the region around Auca Mahuida, as well as prospecting for fossils in rock layers around the city of Neuquén. The rocks throughout the area were spectacularly beautiful, although soaring vultures and condors often kept an ominous eye on our activities. During one of these prospecting side trips about twenty miles from camp, our pickup truck broke down, and the circling vultures seemed to bode ill. Fortunately, Rodolfo had come along with us in his truck, so all ten of us crammed in the cab and the back of his pickup for the rough ride back to camp. It took several days to recover the abandoned truck from the rocky ravines and get it back to Plaza Huincul, where it could be fixed, which left us short on vehicles during the final push to finish our work.

Finally, on November 27, with our contingent of vehicles back at full strength, we broke camp and began the two-day drive back to Buenos Aires. The thrill of discovery still filled our thoughts during the drive, although these thoughts were occasionally interrupted by the buzz of large black bees flying around the interior of the van. The bees belonged to Osvaldo Di Iorio. Osvaldo is an accomplished entomologist, currently working for Argentina's National Council of Science. Throughout his passionate life devoted to the study of insects, Osvaldo had amassed an enormous collection of insects from sites all over Argentina. He has accumulated more than 1 million specimens to date—a collection that rivals that of many museums. Osvaldo had exhibited a willing enthusiasm for quarrying and collecting fossils; however, he had joined us primarily to add to his collection of insects from the dry desert landscapes of this remote area of Patagonia. Accordingly, he would often momentarily suspend his work in the quarry to chase down a bug with the cyanide jar that he always carried in his pocket. Near our camp, Osvaldo had found a large log that contained dozens of holes bored by bees for their larvae. Unbeknownst to the rest of us, he had packed the log in the van to get the larvae back home for his collection. But he hadn't realized that they would hatch the next day on our way home, sending us scurrying to open the windows and shoo them out across the Pampas.

When we safely made it back to Buenos Aires, we were exhausted. All the scientific research required to figure out exactly what we had found still lay before us, and we were eager to get all our fossils back to the lab, where we could prepare the eggs and look for clues to determine what kind of dinosaur had laid them.

CHAPTER FOUR

Compiling a List
of Possible Victims

A Brief History of Dinosaurs

We arrived back in New York during the first week of December 1997. As detailed in our collecting agreements, Rodolfo and his museum in Plaza Huincul gave us permission to borrow several of the fossils we had collected so that Marilyn could prepare them at the Peabody Museum at Yale University and Luis could study them at the American Museum of Natural History. At the same time, Rodolfo and Sergio planned to prepare some of the other blocks of eggs at the Carmen Funes Museum. Our initial task was to compile and whittle down the list of possible victims.

As mentioned earlier, large, round eggs such as the ones we had found in Patagonia had been often identified as belonging to sauropods, and those from the end of the Cretaceous were usually attributed to titanosaurs. This preliminary identification had been based on circumstantial evidence, including the large size of the eggs, the occurrence of titanosaur fossils in the same rock layers as the eggs, and that these types of eggs are found only in deposits that contain skeletal remains of titanosaurs. However, sauropod dinosaurs had not previously been discovered inside any of the round eggs attributed to these dinosaurs, so we could not be certain of this identification. In fact, because of this very reason we could not be sure that sauropods laid eggs at all, an uncertainty that led some paleontologists to speculate that sauropods gave birth to live young.

Dinosaurs are a diverse group of animals. They dominated the continents for more than 150 million years, during which time dozens of groups arose and went extinct. Just about any large dinosaur that lived near the end of the Mesozoic had to be considered as a possible victim of the ancient catastrophe at Auca Mahuevo. The quickest way to compile our list of likely victims was to look back through the evolutionary history of dinosaurs and identify which groups lived in Patagonia at the end of the Mesozoic. This narrative will also provide those interested with a brief review of dinosaur evolution and the method that paleontologists use to reconstruct it.

Since the naming of the first dinosaur in 1824, several hundred kinds of dinosaurs have been discovered, and more are being found all the time. A primary goal of paleontology is to understand how different species are related to one another, in other words, to reconstruct the history of how they evolved. This pursuit is somewhat like studying the genealogy of a family tree. Contemporary paleontologists use a scientific method called cladistics to reconstruct the evolutionary history of extinct groups and draw family trees that link groups of ancient animals and plants. Willi Hennig, a German entomologist, introduced this analytical method in the 1950s, and subsequent researchers have refined it over the last forty years.

Cladistics is based on a fairly simple concept. Although life is diverse, we see a pattern in that diversity when we look for characteristics that are shared by different organisms. This pattern of characteristics can be used to arrange organisms into smaller groups contained within larger groups. The arrangement of groups within groups results from evolution, when descendants inherit new characteristics from their ancestors. By studying how these characteristics are distributed among different animals and plants, we can determine the order in which different groups evolved and thereby interpret the sequence of evolutionary history.

The evolutionary relationships among different groups of organisms can be shown on branching diagrams called cladograms. A clade is simply a group of organisms that includes the first member of the group, or the group's common ancestor, and all of its descendants. In essence, branches on the tree represent different clades of animals or plants, and the branching points on the tree represent common ancestors that possessed new evolutionary characteristics, which were inherited by the descendants of the common ancestor on higher branches of the tree.

For example, some characteristics are shared by a large number of animals. Fish, frogs, dinosaurs, and humans all have a backbone composed of vertebrae. Thus, they all belong to the group of animals called vertebrates, which constitutes a major limb on the family tree of animals. The backbone is thought to have evolved in the very first vertebrate, or the common ancestor of the group. Then, all of its descendants, including humans, inherited a backbone from that common ancestor. Other characteristics are shared by a smaller number of animals within the vertebrate group. For instance, frogs, dinosaurs, and humans have four limbs (arms and legs) with bony wrists, ankles, fingers, and toes, so these animals belong to a subgroup of vertebrates called tetrapods, which means "four-footed." Tetrapods represent a smaller branch on the limb of the tree that contains all the vertebrates. Again, four limbs originally evolved in the ancestor of all tetrapods, and all the descendants of that common ancestor, including humans, inherited some version of its four limbs.

Since having a backbone is more widespread among animals than having four limbs—for example, since fish have backbones but lack limbs—the backbone is thought to have evolved before the limbs did.

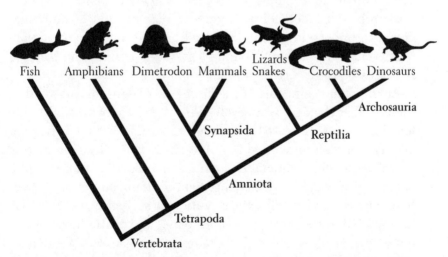

Genealogical relationships of the major groups of vertebrate animals. Birds are included within dinosaurs.

This is why tetrapods are considered to represent a subgroup within vertebrates. In this example, the fossil record appears to confirm this idea, because the first known vertebrate lived about 500 million years ago, whereas the first known tetrapod appeared about 350 million years ago.

However, the fossil record does not always confirm the evolutionary sequence of characteristics suggested by looking at features shared by larger and smaller groups. This is not too surprising because only a small fraction of the organisms that once inhabited the earth are preserved as fossils, and many of those are yet to be discovered. How else could we explain why paleontologists discover new species of extinct dinosaurs and other organisms every year? Because the fossil record is so incomplete, most contemporary paleontologists prefer not to rely on the age of the fossils when reconstructing the family trees of ancient organisms. This task is more accurately achieved by comparing only the physical characteristics, such as backbones and limbs, preserved in the fossils.

Reconstructing the genealogy of plants and animals is often further complicated since the enormous array of characteristics found in these organisms—from physical features to molecular genes—did not evolve only once in a linear fashion. Similar characteristics have evolved in animals that are apparently not closely related. Among living animals, for example, wings evolved in insects, bats, and birds. A warm-blooded metabolism is found in mammals, birds, and tuna. Humans, birds, and kangaroos are all bipedal, meaning that they walk exclusively on their two hind legs. Other examples of features shared by distantly related groups abound.

An important component of cladistics involves sorting out whether these similar characteristics evolved from the same ancestor or different ancestors. To make this decision scientists rely on the principle of parsimony, often referred to as Occam's razor. Given alternative genealogical interpretations, the principle chooses the simplest explanation that accounts for the arrangement of groups on the family tree. In other words, it identifies the evolutionary sequence of events that is supported by the most pieces of evidence and contradicted by the fewest pieces of evidence. For example, there are two alternative interpretations involving the genealogy of bats: Are bats more closely related to birds or to mammals? Although bats and birds both have wings and are warm-blooded, a greater number of characteristics provide evidence that bats are more closely related to other mammals than they are to birds. Bats have hair, nurse their young with milk, have three bones inside their ears, and have teeth of different shapes like other mammals. Birds do not share these characteristics, and in fact, the wings of birds and bats are constructed quite differently. Whereas five fingers of the hand support the membrane that forms a bat's wing, the feathers of a bird's wing are attached to the arm and only one finger of the hand. This difference is interesting, but the number of similarities shared by bats and mammals is more important for interpreting the genealogy of bats. Using the principle of parsimony, we can see that most of the anatomical evidence suggests that bats are more closely related to mammals. Consequently, birds and bats are not as closely related, even though they both have wings and are warm-blooded. In the end, cladistic methodology compares only similarities shared among different groups, then uses the principle of parsimony to choose the genealogical interpretation that is supported by the larger number of similarities.

The previous example may seem rather obvious, since most biology students have learned that bats are mammals and birds are not, but cladistics and the principle of parsimony have shed new light on other evolutionary puzzles that had previously proved more difficult to solve. One of these is the evolutionary relationship of birds to crocodiles, lizards, and other reptiles. At first glance, crocodiles look much more like lizards than birds. Crocodiles and lizards have a scaly, reptilelike body, whereas birds have a feathered body with wings. Yet, despite the obvious differences seen in crocodiles and birds, a closer examination reveals that crocodiles and birds share many more physical similarities than crocodiles and lizards do. For example, crocodiles and birds share a series of air holes in their middle ear that are not found in lizards. Birds and crocodiles also share a robust rib cage that is strengthened by short struts of cartilage called uncinate processes. In birds and some theropod dinosaurs, these uncinate processes ossify. Another feature found in crocodiles and birds is a muscular stomach or gizzard that processes food. Lizards do not have gizzards. Consequently, when we apply the principle of parsimony, crocodiles and birds are more closely related than crocodiles and lizards on the family tree of vertebrates. The similar reptilian appearance of lizards and crocodiles is evolutionarily misleading and is refuted by the careful scrutiny inherent in cladistics; so most evolutionary biologists now classify crocodiles and birds in the same evolutionary group (Archosauria), which is a subset of lizards and all other reptiles.

Understanding the genealogical relationships between ancient organisms is crucial for reconstructing the origin of living animals and plants, as well as for understanding the evolution of their anatomical, physiological, and behavioral systems. When looking at a particular characteristic, such as our grasping hand, the explanation of its evolutionary origin would be very different if we assumed that humans originated from bats rather than primates. If we originated from bats, our grasping hand would have had to evolve from a structure used for flying, which would be difficult to imagine, although not impossible. However, since many other characteristics of our skeleton indicate that humans evolved from other primates, which use their hands for holding on to branches and gathering food, the evolutionary origin of our grasping hands is easier to understand.

With this concept for reconstructing evolutionary history in mind, we can trace some of the steps that led to the evolution of the major groups of dinosaurs by working our way up the branches and branching points of the dinosaur family tree. The earliest known dinosaurs appear in the fossil record about 230 million years ago. Some of them, such as the herbivorous *Pisanosaurus*, and the carnivorous *Herrerasaurus* and *Eoraptor*, were found in northwest Argentina. The fact that these dinosaurs already had specialized anatomical features for eating plants and flesh, which distinguish them as members of specific dinosaur groups, indicates that there must have been earlier dinosaurs, although we have yet to find fossils of them. The first question that arises in looking at these and other dinosaurs is, what makes a dinosaur a dinosaur? In other words, what characteristic evolved in the common ancestor or very first dinosaur? To find out, we need to search for a characteristic that is found in all dinosaurs but not found in other reptiles, including turtles, lizards, snakes, and crocodiles.

The story of the evolution of dinosaurs basically revolves around locomotion. The common ancestor of dinosaurs possessed a hip structure different from that found in other reptiles. All dinosaurs have a hip socket, or acetabulum, that had a hole in the middle of it and a strongly developed ridge of bone along the top of the socket. As a result, dinosaurs inherited a vastly different posture and gait from their common ancestor.

To illustrate this, think of the way a lizard stands. Its hind limbs extend horizontally out from its hips before the lower hind limb bones reach vertically down to the ground. The limbs support the rest of the body in a sprawling posture, and when the lizard moves, its body basically makes an S-shaped motion. The acetabulum in a lizard is solid, with no hole in the middle, which makes sense structurally. Where the lizard's thighbone, or femur, meets the hip, a lot of force is generated by the muscles that pull the bone horizontally into the hip socket. The acetabulum is solid to help resist those horizontally directed forces, and no extra bone is needed along the upper margin to help provide support.

In contrast, dinosaurs have hind limbs that extend vertically down from the hips to the ground, resulting in a more upright or erect posture. Clear evidence for this is found in sequences of fossil foot-

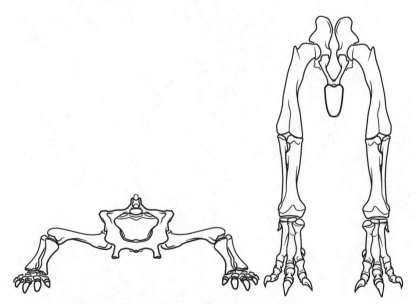

Figure contrasting sprawling posture in a turtle (left) with erect posture in a dinosaur (right).

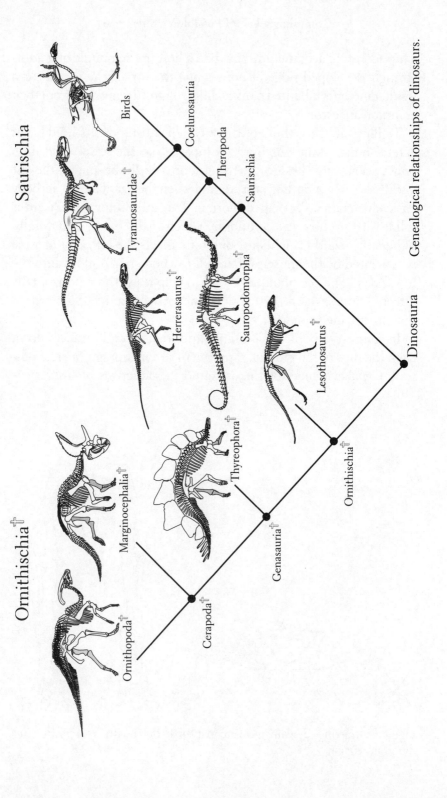

Genealogical relationships of dinosaurs.

prints called trackways, which show that when dinosaurs walked, the left and right footprints form a nearly straight line with one another. With this arrangement of the hind limbs, the femur exerted force in a much different direction where it fit into the acetabulum. The force generated by the dinosaur's weight was directed toward the upper margin of the hip socket rather than toward the center, with the bony ridge at the top of the acetabulum helping to counteract that force. And because no force was directed toward the center of the socket, a hole is found there. Dinosaurs were, thus, the first reptiles in which a fully erect stance evolved.

Two major groups within dinosaurs represent the two large limbs on the family tree of dinosaurs. One is called Ornithischia, which contains most of the groups of herbivorous dinosaurs, including armored dinosaurs, duckbills, horned dinosaurs, and dome-headed dinosaurs. The 230-million-year-old *Pisanosaurus* is the oldest known ornithischian dinosaur, and like all other early dinosaurs, *Pisanosaurus* walked on its two hind legs. This suggests that the common ancestor of dinosaurs must also have been bipedal. Even at this early date in dinosaur evolution, the plant-eating *Pisanosaurus* already exhibits the physical characteristics that are typically found in the group. Ornithischians evolved from a common ancestor in which one of its hipbones, called the pubis, points toward the rear of the animal. In the common ancestor of all dinosaurs, the pubis pointed forward. It is not immediately obvious what purpose this change in direction served. Perhaps it provided extra support for the gut, which had to be large to process the enormous amount of vegetation that was required to nourish the herbivorous ornithischians, but no one knows for sure.

Armored dinosaurs represent one of the most spectacular branches on the ornithischian limb of the dinosaur family tree. Noteworthy members include the tanklike ankylosaurs and the spectacularly ornamented stegosaurs. Together, these two lineages make up the group called thyreophorans, which evolved from a common ancestor with a mosaic of bony armor that essentially covered the entire body. Even the earliest thyreophorans walked on all four legs, probably to better support the weight of their heavy, bulky body. The armor, which often included enormous spikes and plates, almost certainly played a role in protecting these dinosaurs from the predatory dinosaurs that roamed the environment. However, the armor may also have played a role in helping these dinosaurs recognize members of their own species and potential mates. Because we cannot observe the behavior of these animals in the wild, again, we simply cannot be certain.

The duckbills, horned dinosaurs, and dome-headed dinosaurs comprise a group called Cerapoda. Unlike the thyreophorans, most of these dinosaurs retained the hind-legged gait of their early ancestors. Only within the horned dinosaurs did a four-legged gait evolve, illustrating a common theme that often recurred in the evolution of large, heavy groups of dinosaurs. Cerapods evolved from a common ancestor that had an uneven covering of enamel on the inside and outside surfaces of their teeth, which, like other reptiles, they replaced continually throughout their life. In many cerapods, a complex mosaic of teeth grew one on top of the other in the jaws. Presumably, the uneven covering of enamel helped to maintain a rough, rasplike surface on top of the batteries of teeth to shred and crush the vegetation that they ate.

Duckbills, which are also called hadrosaurs, represent an extremely diverse and abundant lineage on the family tree of cerapods. They arose from a common ancestor in which the hingelike joint between the upper and lower jaws lay below the level of the tooth rows. This arrangement resulted in an extremely powerful crushing action as the chewing musculature clamped the jaws together. The lower position of the jaw joint, along with the rasplike chewing surface on top of the tooth rows, provided duckbills with an ability to grind up tough vegetation that was unmatched in other groups of dinosaurs. Undoubtedly, this specialized feeding adaptation played an important role in the evolutionary success of duckbills. In North America, these dinosaurs are among the most commonly found. Their fossil remains are sometimes preserved in large accumulations known as bone beds, which presumably represent death assemblages caused by the catastrophic demise of huge herds that contained individuals of all ages. Large herds of duckbills probably roamed the floodplains along major rivers that ran across the continent toward the end of the Mesozoic.

In spite of their distinctive jaw apparatus, however, duckbills are most commonly recognized by the elaborate bony crests that adorned the top of the skull in many species. The function of these crests has intrigued paleontologists for more than a century. Some have speculated that the large air passages contained within the crests served as resonating chambers to amplify vocal calls made by the duckbills, while other paleontologists have suggested that the crests were used to identify other members of the same species and potential mates. Again, it is difficult to be certain since all the duckbills are extinct, and perhaps the crests were involved in both these activities.

The horned dinosaurs and dome-headed dinosaurs form a group within cerapods called the Marginocephalia. The name comes from the shelf of bone rimming the back of the skull, which was inherited from the common ancestor of the group. Evolution took this bony shelf and expanded it in two different directions in the horned dinosaurs and dome-headed dinosaurs. In horned dinosaurs, also known as ceratopsians, the bony shelf at the back of the skull expanded backward to form a shieldlike structure called a frill. Although the frill is relatively small in more primitive members of the group, such as the parrotlike *Psittacosaurus* and its distant relative *Protoceratops*, it became a large and elaborate structure in many later and larger members of the group, such as *Triceratops* and *Styracosaurus*. In dome-headed dinosaurs, also known as pachycephalosaurs, the bony shelf along the back of the skull expanded forward to form a thick, bony helmet on top of the skull. In some later members of the group, such as *Pachycephalosaurus*, this dome is about six inches thick. Furthermore, while horned dinosaurs developed a four-legged gait in the more advanced members of the group, such as *Triceratops*, the dome-headed dinosaurs retained the two-legged stance and gait of the earliest dinosaurs.

The second major limb at the base of the family tree contains the saurischians, who evolved from a common ancestor that had a hand capable of grasping. The thumb was slightly offset from the rest of the fingers. The bony structure of the saurischian's grasping hand differs slightly from the structure of our hand, but the basic result was the same. Within saurischians, we find two large branches on the evolutionary tree. One contains all the giant, long-necked, four-legged herbivores called sauropods. The other contains all the two-legged, carnivorous theropods, including their descendants, birds.

Theropods evolved from a common ancestor that had only three fully developed toes on the hind feet, with the central toe being the longest. The common ancestor of theropods gave rise to many

Genealogical relationships of meat-eating theropod dinosaurs.

branches near the base of this limb on the dinosaur family tree. For the most part, these branches contain relatively primitive, meat-eating dinosaurs. Some of these are among the oldest dinosaurs that we know about, such as *Herrerasaurus* and *Eoraptor,* which were collected from rocks in northwest Argentina that are about 230 million years old. Another branch near the base of the limb represents the abelisaurs, including the Argentine *Carnotaurus.* This fearsome predator lived later in the Mesozoic era, about 75–80 million years ago, although the lineage leading to this imposing branch of dinosaurs must extend back millions of years before this. As mentioned earlier, *Carnotaurus* was unusual in that it had short arms and sported two prominent bony horns above its eyes. We will delve into more details regarding abelisaurs later in our story.

The other larger branch of theropods contains a group called tetanurans. These dinosaurs evolved from a common ancestor that possessed collarbones that were fused together like the wishbone of birds, as well as a hand that had no more than three fingers. Branches near the base of the tetanuran limb of the tree include the giant meat-eaters *Allosaurus* and *Giganotosaurus. Giganotosaurus,* one of the largest of all the known carnivorous dinosaurs, had been found in the same province of Argentina where our expedition conducted its exploration, and our colleague Rodolfo had been instrumental in collecting and describing the massive skeleton.

Another branch within tetanurans contains smaller meat-eaters, such as *Velociraptor* and *Compsognathus* of *Jurassic Park* fame, as well as birds and the infamous *Tyrannosaurus.* This group contains all dinosaurs called coelurosaurs, who evolved from a common ancestor that had an extra hole in the snout for reducing the weight of its skull and an elongated ilium, the main bone that forms the top part of the hip. The first coelurosaur also had elongated arms, although this feature was later modified in many of its descendants, including *Tyrannosaurus* and the kiwi. For those descendants that retained long arms, these appendages may well have been an advantage in capturing prey.

Spectacular new discoveries in China have shown that primitive coelurosaurs were covered with downy structures that are interpreted to be the evolutionary precursors of the vaned feathers found in birds. These discoveries suggest that most, if not all, coelurosaurs

The Patagonian theropod *Giganotosaurus* is among the largest carnivorous dinosaurs. Our crew discovered the remains of some of its relatives at Auca Mahuevo.

were feathered; even the colossal *Tyrannosaurus* must have had a feathered body at some early stage of its life. However, scientists do not believe that adult tyrannosaurs were feathered because the combination of this insulating covering and their large size might have posed a disadvantage in regulating the animal's body temperature.

One branch within coelurosaurs leads to the "bird-mimic" dinosaurs called ornithomimids. Ornithomimids have a skeleton that looks superficially like that of an ostrich, which explains the derivation of their name. That these dinosaurs had long legs with shortened toes, which would have reduced the amount of friction with the ground, suggests that ornithomimids were swift runners. Most ornithomimids lack teeth in their jaws, which has led to some questions about what these animals ate. Recent discoveries of stones called gastroliths inside the stomach area of some ornithomimid skeletons suggest that these dinosaurs may have eaten plants because similar stones are commonly found in the gizzard of herbivorous birds and other plant-eating dinosaurs. Yet gastroliths are also known from other theropods that clearly had a carnivorous diet.

The other branch within coelurosaurs leads to birds and small, meat-eating dinosaurs, such as *Velociraptor*. These evolved from a common ancestor that possessed a crescent-shaped bone in their wrist called the semilunate, because it is shaped like a crescent moon, and hips in which the pubic bone pointed toward the rear of the animal. The shape of the semilunate bone allows the wings to be folded back against the body in the distinctive position that we see in birds today. The presence of this bone in *Velociraptor* and its relatives indicates that these dinosaurs were also able to fold their arms back against their body in the same way that birds do. These dinosaurs form the group called maniraptors, named for their extremely enlarged and elongated hands, which were certainly formidable predatory weapons.

As odd as it may seem, birds are maniraptors and, therefore, dinosaurs. They belong on the same branch of the dinosaur family tree that contains *Velociraptor*. Earlier, when we introduced the fundamentals of cladistic methodology and the principle of parsimony, we downplayed the role of physical differences and emphasized the role of similarities in reconstructing the evolutionary relationships of organisms. The same concepts must be applied to reconstruct the evolutionary relationships between birds and other dinosaurs.

Superficially, *Velociraptor* and other well-known maniraptors may appear very different from living birds. While the fearsome star of *Jurassic Park* possessed large claws and sharp teeth, birds lack both these features. Yet, a closer examination of the skeletons in birds and *Velociraptor* reveals numerous similarities, and these become even clearer when one compares the skeleton of *Velociraptor* with those of ancient Mesozoic birds.

Birds not only share the unique structure of the hips and wrists found in other maniraptors, but also the long arms of coelurosaurs, the three-fingered hands of tetanurans, the three-toed feet of theropods, and the perforated hip socket of dinosaurs. In addition, they have feathers, like their most immediate forerunners among the maniraptors and coelurosaurs. In the last few years, spectacular discoveries from the northeastern corner of China have shown that the body of maniraptors was also covered with feathers. A close cousin of *Velociraptor*, the Chinese maniraptor called *Sinornithosaurus* had two-to-three-inch-long downy feathers covering its skin. The fact that feathers, long thought to be a characteristic unique to birds, have been found on fossil skeletons of other maniraptoran dinosaurs has dealt the final blow to paleontologists who doubted that birds evolved from dinosaurs. Today, we can declare that birds are dinosaurs with the same degree of confidence that we can say that humans are primates.

Birds experienced a long and complex evolutionary history, most of which was played out in the Mesozoic era. The earliest known bird is *Archaeopteryx*, which lived about 150 million years ago in the late Jurassic period. Exquisite fossils of this crow-sized bird were first found in the limestone quarries of southern Germany in the mid-1800s. The skeleton of *Archaeopteryx* still had a very dinosaurian appearance, with a long tail, sharp teeth, and powerful claws on its wings, although its plumage was fully modern. Within 20 million years after *Archaeopteryx* lived, birds had already evolved with a more typical, stumpy tail and wings that gave them a more modern appearance, though birds would retain their teeth throughout the Mesozoic.

Another early branch on the family tree of birds contains all Enantiornithes, who were sophisticated fliers, even though the first known members of the group lived as early as 125 million years ago. It is rather amazing to realize that even that far back in time, birds were able to perform the same aerodynamic feats that they delight us with today. The common ancestor of Enantiornithes and modern birds possessed a small tuft of feathers that attaches to the first finger, or thumb, of the hand. This "bastard wing" was an important aerodynamic innovation and somewhat comparable to the flaps of an airplane. The fine-tuning of flight capabilities led to a large radiation of birds. Enantiornithes quickly adapted to life in the water, in the air, across more open terrain, and even in deserts. In fact, the primary initial goal of our expedition was to find fossils of this primitive group of birds.

The other major branch within saurischians includes the giant plant-eating dinosaurs called sauropodomorphs, which arose from a common ancestor with a relatively long neck and small head. This group contains the largest animals ever to have walked on land, and walk they did. Their remains have been recovered from every continent.

The earliest member of the group is *Saturnalia,* a 230-million-year-old dinosaur from the late Triassic of Brazil, who had a gracile body that was about five feet long with a small head and a long neck. A more advanced and better known primitive sauropodomorph is the

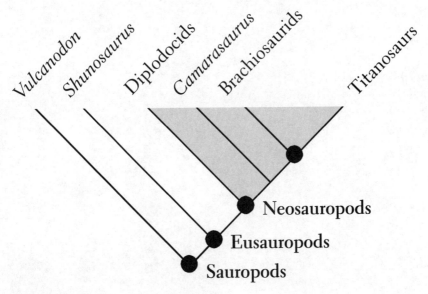

Genealogical relationships of sauropods.

The dicraeosaurid sauropod *Amargasaurus* lived in Patagonia during the early Cretaceous period. Its back was lined with two rows of tall spines that projected from each of its backbones.

fifteen-foot-long *Plateosaurus*, which possessed a grasping hand that it inherited from the common ancestor of all saurischians. Its hind limbs were much longer and stronger than its arms, and it is clear from the proportions of these limbs that *Plateosaurus*, like the smaller *Saturnalia*, walked only on its hind legs.

The main branch on the sauropodomorph limb of the dinosaur's family tree contains all the enormous sauropods. In these animals, all four limbs were strongly developed like the columns of a building, indicating that they walked on four legs. The bones of their wrists and ankles were greatly reduced in number, presumably an adaptation for supporting their heavy weight. Fossilized trackways confirm their quadrupedal mode of locomotion. The four-legged gait of sauropods represents another example of the recurrent evolution of this type of locomotion in large dinosaurs, probably to help support the weight of their colossal bodies.

Within sauropods, there are several branches on the evolutionary tree, although scientists are still debating how the different groups are related to one another. These groups include diplodocids, dicraeosaurids, camarasaurids, brachiosaurids, and titanosaurs. Diplodocids are exemplified by *Diplodocus*, the extremely long-necked, whip-tailed sauropod of the late Jurassic in North America. Although one of the longest dinosaurs ever discovered, it probably did not weigh as much as some of its more robust sauropod relatives. Its teeth are elongated, peglike or pencil-like structures that were probably used to strip vegetation off branches.

The dicraeosaurids include *Dicraeosaurus*, a relatively small, short-necked sauropod from the late Jurassic of Tanzania, and *Amargasaurus*, a Patagonian form that lived in the early Cretaceous. Dicraeosaurids evolved from a common ancestor with tall spines on top of its vertebrae, especially in the hip region, and peglike teeth similar to those of diplodocids. Most paleontologists agree that dicraeosaurids are very closely related to diplodocids and believe that their peglike teeth evolved in the common ancestor of these two groups.

Brachiosaurids have longer front legs than hind legs, a posture present in their common ancestor that made them some of the tallest dinosaurs ever found. The immense *Brachiosaurus* was discovered in the late Jurassic of Tanzania by a German expedition at

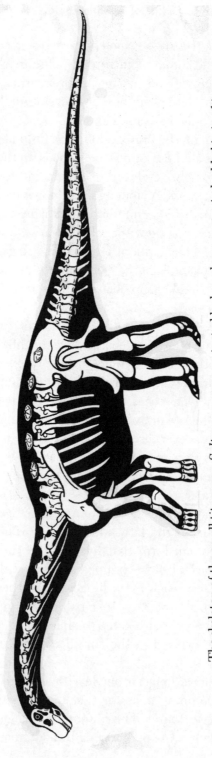

The skeleton of the small titanosaur *Saltasaurus* was protected by large bony scutes imbedded in its skin.

the beginning of the twentieth century. This dinosaur is probably the tallest of all, with an extremely long neck extending above its tall shoulders.

Titanosaurs range from medium to gigantic in size. The titanosaur *Argentinosaurus*, with its 100-ton, 120-foot-long body, is the largest dinosaur ever found. It may well have shaken the Patagonian landscape with every step it took. Titanosaurs are the most diverse group of sauropods; their skeletons are known from South America, North America, Africa, Europe, and Asia. Most lived during the Cretaceous period, although the earliest known form is *Janenschia*, from the late Jurassic of Tanzania, and more than a dozen species inhabited South America. Titanosaurs evolved from a common ancestor that had small oval air holes in their vertebrae, an important adaptation for lightening their heavy skeletons. There were as many as thirteen vertebrae in the neck, twelve in the trunk, six in the sacrum, and more than thirty in the highly flexible tail. The spines on top of the neck and trunk vertebrae were simple, instead of being split as in diplodocids. Advanced titanosaurs had tail vertebrae with a concave front surface, and their teeth resemble the pencil-like teeth of diplodocids. The front legs of these animals were shorter than the hind legs. One of their most remarkable characteristics is the bony covering of armor formed by thousands of small, rounded lumps and a few larger plates, although it is unclear that this formidable armor developed in all members of the titanosaur lineage.

This abbreviated cladogram of dinosaurs provided us with a list of potential victims that could have perished inside the dinosaur eggs at Auca Mahuevo. Our next job was to begin eliminating candidates by determining what kinds of dinosaurs had previously been found in Patagonian rocks, especially ones that were from about the same age as the dinosaurs collected at Auca Mahuevo.

Many diverse lineages of dinosaurs have been discovered in Patagonia's vast Mesozoic rock layers, extending from the late Triassic up to the very end of the Cretaceous, a period of more than 130 million years. In the past, only a handful of discoveries of ornithischians had been made, but recent findings are demonstrating that the history of ornithischians in Patagonia is much richer than previously thought. All Patagonian ornithischians are restricted to the Cretaceous. Although specimens of ceratopsians, stegosaurs, and ankylosaurs are known from

With an enormous body supported by its pillar-like limbs and backbones of higher than 5 feet, the Patagonian titanosaur *Argentinosaurus* could have reached lengths of 120 feet.

fragmentary remains, duckbills and their primitive relatives are much better represented. Both primitive duckbills and their primitive forerunners are known from late Cretaceous rocks of northern Patagonia. *Anabisetia* and *Gasparinisaura* are two small, distant cousins of the duckbills known from Neuquén and Río Negro provinces, respectively. True duckbills, known from the late Cretaceous of northern Patagonia, include *Secernosaurus*, *Kritosaurus*, and another unnamed crested form. In contrast to their North American relatives, duckbills appear to have been less successful in South America, where late Cretaceous ecosystems were dominated by huge, herbivorous sauropods.

The earliest known Patagonian theropod is from the early middle Jurassic of Chubut, but it is very fragmentary. Much more complete is *Piatnitzkysaurus* from the middle Jurassic of Chubut. *Piatnitzkysaurus* was a moderately sized, meat-eating, primitive tetanuran. Many more Cretaceous theropods have come from Patagonia, the most abundant of which are the abelisaurs. An early member of this group is the early Cretaceous *Ligabueino*, an animal the size of a pigeon. Much larger abelisaurs lived in the late Cretaceous of Patagonia, including the horned *Carnotaurus*, one of the largest carnivores of its time.

A diverse assemblage of sauropodomorphs has also been found in Patagonia. In the Triassic, prosauropods roamed the land and nested there. Several specimens of these, including adults, juveniles, hatchlings, and eggs, were collected by José Bonaparte in the late Triassic rocks of Santa Cruz province, in the southern end of Patagonia. The spectacularly preserved hatchlings, known by the name of *Mussaurus*, which means "mouselike lizard," were little more than six inches long. The earliest known sauropod from Patagonia is *Amygalodon*, from the early part of the middle Jurassic, which is known from isolated teeth and a vertebra that were collected in the province of Chubut. Bonaparte and his collaborators found better sauropod remains in the middle Jurassic of Chubut in the late 1970s. An incomplete skeleton constitutes the only known specimen of the dinosaur *Volkheimeria*. Another middle Jurassic sauropod from Patagonia is *Patagosaurus*, which Bonaparte discovered in the same rock layers that produced *Volkheimeria* and the theropod *Piatnitzkysaurus*. The primitive *Patagosaurus* is known from several specimens of both adults and juveniles.

Abundant remains of sauropods are also known from the Cretaceous of Patagonia. A peculiar one from the early Cretaceous is the dicraeosaur *Amargasaurus*, a close relative of the late Jurassic *Dicraeosaurus* from Africa, unearthed from the same strata as the tiny theropod *Ligabueino*. One nearly complete skeleton was found in which most of the animal's bones were still articulated. These dinosaurs have a double row of long spines that project upward from their backbones, as well as peglike teeth like those found in their cousins the diplodocids. Another relative of the whip-tailed diplodocids is *Rebachisaurus*, a middle Cretaceous sauropod that is also known from Africa.

Titanosaurs are the most common sauropods from the late Cretaceous of Patagonia. As we've seen, these were the first dinosaurs to be found on the South American continent. Numerous remains of these dinosaurs have been found in the northern half of the region, particularly in rocks within the province of Neuquén, where we were searching. In addition to *Argentinosaurus*, other Patagonian titanosaurids include *Andesaurus*, *Epachthosaurus*, *Aeolosaurus*, *Neuquensaurus*, and several other species.

The list of candidates that could have laid the eggs at Auca Mahuevo was extensive. So we began by paring it down to those that came from late-Cretaceous rocks of Patagonia, since we knew from previous studies that the rocks we were collecting were from this age. The embryos from Auca Mahuevo could have belonged to dinosaur groups that lived millions of years earlier or on other continents, but to start, our list of the most likely candidates included only the dinosaurs known to have lived in the region at that time. Possible candidates among ornithischians included duckbills and their forerunners; among theropods, the abelisaurs were candidates; and among sauropods, titanosaurids constituted likely victims.

We were now ready to begin to search for more definitive clues to help us solve the mystery. The bony evidence we needed to make a positive identification of the victims at Auca Mahuevo was hidden inside the eggs.

CHAPTER FIVE

Dinosaur Eggs

Evolutionary Time Capsules

Dinosaur eggs have been found on all continents except Antarctica. Among the oldest are those associated with the tiny prosauropod hatchlings of *Mussasaurus* from the late Triassic of South America. The few known late Jurassic sites are restricted to North America, Europe, and Tanzania. Eggs from the early Cretaceous are far more abundant, but with the exception of one site from southern Australia, all are restricted to the Northern Hemisphere. Although a large number of late Cretaceous sites yielding dinosaur eggs have been found across North America, Europe, and Asia, relatively few have been discovered in the Southern Hemisphere. Only South America and India have provided evidence of dinosaur nesting sites in the southern continents at the end of the Mesozoic, probably due to the limited amount of paleontological exploration of the continents that occupied the Southern Hemisphere at that time (including India).

An egg for an embryo is like a space suit for an astronaut. A dinosaur egg has numerous special structures to nourish the embryo growing inside and provide protection from the hostile environment outside. The most obvious is the shell. A few reptiles, such as sea turtles, some lizards, and snakes, lay flexible or soft-shelled eggs, but in dinosaurs, as well as in most other reptiles, the shell is primarily composed of hard mineral crystals called calcium carbonate—the same basic material that makes up limestone and cement. These crystals, which are arranged into distinctive structural units that fit tightly together and occasionally interlock, form a hard shell that creates a formidable barrier against bacteria, fungi, and other organisms that can cause disease. However, the shell cannot be too hard, or the embryo cannot break out of it when it is ready to hatch. In actuality, even hard eggshell is quite porous, with microscopic holes that allow gases and water vapor to pass in and out. Oxygen penetrates through the pores into the egg so the embryo can breathe, and carbon dioxide passes out through the pores into the atmosphere.

Inside the egg, a flexible container composed primarily of proteins and fat called the yolk sac (the yellow part of a chicken egg) contains food for the growing embryo and antibodies to help protect the embryo from disease. Another membranous sac, the allantois, serves as a receptacle for waste products. The yolk and allantois are surrounded by albumen, a water-saturated gel that absorbs whatever shocks might jostle the embryo developing at the center of the yolk. The albumen also contains more chemical components to help fend off dangerous microbes. All these structures work together to keep the embryo at a constant temperature inside a fluid environment that cushions it from the extremes and threats of the world outside.

Eggs are typically classified based on visible characteristics of the egg and the structure of its shell, including size, shape, distribution

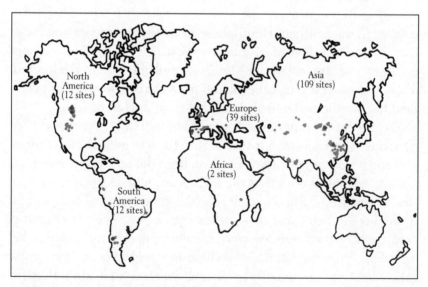

Main sites where dinosaur eggs have been found.

and number of pores, and ornamentation on the shell's surface. At the microscopic level, eggs are classified based on the characteristics of crystalline units that form the shell, the structure of the pore canals, and the chemical composition of the crystalline layers that make up the shell. Such detailed study requires microscopes that analyze polarized light and scanning electron microscopes, also known as SEMs, which magnify the features of the shell tens or hundreds of thousands of times.

Based on the characteristics of the shell units, hard-shelled eggs of living animals are divided into four basic categories: testudoid (some turtles), geckonoid (geckos), crocodiloid (crocodiles), and ornithoid (birds). These eggs come in two main structural types: those made of calcium carbonate crystals called aragonite (some turtles), and those made of calcium carbonate crystals with a slightly different chemical

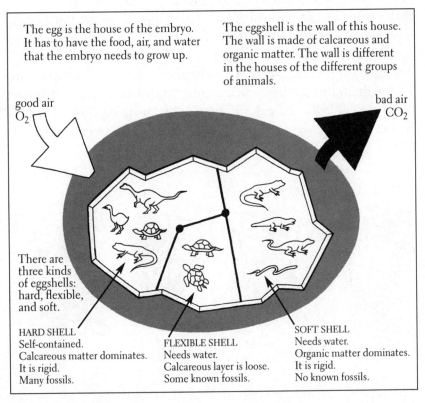

The egg is the house of the embryo. It has to have the food, air, and water that the embryo needs to grow up.

The eggshell is the wall of this house. The wall is made of calcareous and organic matter. The wall is different in the houses of the different groups of animals.

good air
O_2

bad air
CO_2

There are three kinds of eggshells: hard, flexible, and soft.

HARD SHELL
Self-contained.
Calcareous matter dominates.
It is rigid.
Many fossils.

FLEXIBLE SHELL
Needs water.
Calcareous layer is loose.
Some known fossils.

SOFT SHELL
Needs water.
Organic matter dominates.
It is rigid.
No known fossils.

Main functions of an egg.

composition called calcite (crocodiles, geckos, birds, and other dinosaurs).

Some dinosaur eggs, including those of birds, fit into the ornithoid type, but two other categories have been created to accommodate the rest of the dinosaur eggs that have been found. These are the dinosauroid spherulitic and the dinosauroid prismatic types, which are based on the form and structure of the crystalline units that make up the shell, just as in the case of living animals. To see these features, the shell is often sliced to create a thin cross section that cuts through the crystalline units, and the shell structure is then observed through a polarizing microscope, which utilizes light that vibrates in only two planes. French scientists first used this traditional method of slicing the shell into thin sections in the late 1800s, when research on the microscopic structure of fossil eggshell originated.

This approach allows paleontologists to identify the structural type of the shell, but a more comprehensive understanding of the shell's structure can be obtained when this approach is combined with observations made with a SEM. To be viewed with the SEM, the pieces of the eggshell often need to be coated with a thin layer of gold. Some newer SEMs, called environmental SEMs, can observe the shell without this coating, which is great for viewing unique specimens that paleontologists cannot risk coating or damaging, but the resolution of the images is not as good. Another approach is to view the shell

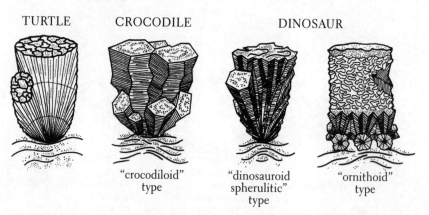

TURTLE   CROCODILE   DINOSAUR

"crocodiloid" type   "dinosauroid spherulitic" type   "ornithoid" type

Turtles, crocodiles, and dinosaurs have specific types of eggshell microstructures. The study of eggshells is an active field of vertebrate paleontology.

from directly above the outer and inner surfaces, rather than in cross section. Both of these approaches provide information about a shell's crystalline structure and pore patterns.

Dinosaur eggs come in many shapes: round like a softball, oval like a football, and elongated like a loaf of French bread. They also range greatly in size, with those of many extinct Mesozoic dinosaurs being as large as that of an ostrich. Interestingly, the dinosaur eggs exhibiting the greatest variation in size are those of the dinosaurs that still live, birds. They can be as tiny as those of a hummingbird or as large as those of an elephant bird, a large, flightless species that lived in Madagascar until a thousand years ago and laid eggs ten times the size of an ostrich egg, much larger than those of any Mesozoic dinosaur.

Our Patagonian eggs are round and relatively large. With an average diameter of five to six inches, they are about the size of a softball and have a volume roughly equivalent to that of a dozen chicken eggs. Today most eggs are preserved in the shape of a disk, probably a consequence of compaction when they were buried beneath thick layers of rock for millions of years, but originally, they were probably almost spherical. Their dark gray surface has a dense ornamentation of small bumps that sometimes coalesce into ridges shaped like worms, but the distinctness of this ornamentation varies widely, probably due to water seeping through the ground and partially dissolving the calcite shell. It is not possible to tell whether the dark gray color we see today was the original color of the egg or whether it resulted from changes that occurred during fossilization. It is also impossible to tell whether the eggs originally had a marbled coloration, typical of many living birds.

Three different structural types of dinosaur eggshell are typically recognized, although the boundaries between these categories are somewhat fuzzy. One widespread type is termed spherulitic, in which an inner core supports stacks of calcite crystals that radiate out from it. This structural type is usually considered typical of all dinosaurs except theropods, and the eggs from Auca Mahuevo are of this type. In the second type, called prismatic, the shell is composed of an inner core that supports two crystalline layers, but the boundary between them is poorly defined. The inner layer is formed by the same radiating crystalline structure typical of the external layer in spherulitic eggshell, while the outer layer has a more prismlike struc-

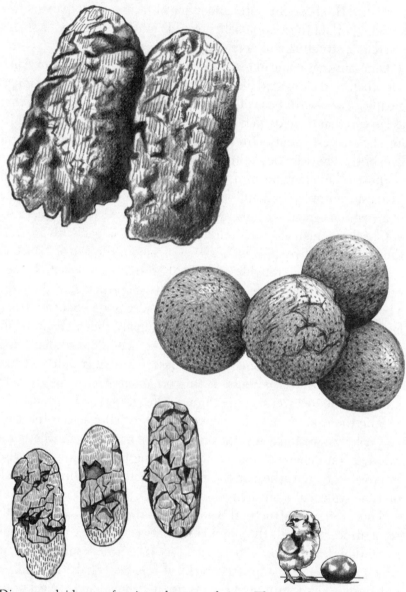

Dinosaurs laid eggs of various shapes and sizes. The round eggs of the Auca Mahuevo sauropods (center right) greatly differ from the elongate eggs of carnivores such as therizinosaurs (top left, about 12 inches long) and oviraptorids (bottom left, about 8 inches long). Bird eggs (bottom right) are also dinosaur eggs.

ture. Prismatic eggshell appears to be typical of primitive theropods. The third structural type, called ornithoid, is also composed of an inner core that supports two or more crystalline zones, but in this case, the boundary between these crystalline zones is abrupt. The lower or inner part of the shell is similar to that of the prismatic type, but the upper or outer part has a scaly appearance. Ornithoid structure is typical of advanced theropods, including all birds.

This threefold classification based on microscopic structure has dominated studies of dinosaur eggshells for several decades, but recent investigations suggest that some revisions are necessary. Just as the bony structures of dinosaur skeletons can be used to classify dinosaurs into different evolutionary groups, so can the microscopic structure of their eggshells. If the cladistic method is used, eggshell structure might provide important new information about how different groups of dinosaurs evolved from one another, since the structure of dinosaur eggs evolved right along with the animals that laid them. Initial research along these lines has suggested that the existing, threefold system used to categorize dinosaur eggs may be hindering our attempts to glean information about dinosaur evolution from the structure of their eggshells.

Paleontologists have, therefore, begun to reanalyze and recategorize different kinds of dinosaur eggshell, and the preliminary results seem promising. For example, all dinosaur eggshells, including those seen in bird eggs, grow outward from an inner core often referred to as the *organic core*, which indicates that all the different kinds of dinosaur eggs evolved from one kind of egg laid by a single common ancestor. In crocodiles, distant relatives of dinosaurs, the inner organic core is poorly differentiated if not absent. This is not surprising, since the structure of the hips in all dinosaurs, including birds, also suggests that dinosaurs evolved from a single common ancestor, as we discussed in an earlier chapter. Furthermore, the presence of two or more distinct structural layers in the eggshells of birds and theropod dinosaurs provides more evidence to suggest that birds evolved from meat-eating theropods. The ornithoid type of eggshell found in living birds is composed of a minimum of two distinct layers. The inner layer is composed of calcite crystals that radiate out from the core, while the outer layer(s) have a more scaly appearance. A similar structural composition is found in some maniraptors, suggesting that birds and maniraptors evolved from the same common ancestor, one that laid eggs with shells divided into two or more layers. Again, as noted earlier, the bony structure of the wrist and other skeletal parts, along with the presence of feathers in some theropods and birds, suggests the same thing. Perhaps the restriction of the radial section of the dinosaur prismatic and ornithoid types to the inner portion of the eggshell is a characteristic that blurs the distinction between these two types. Alternatively, it may be that dinosaurs that laid these two types shared a common evolutionary origin that was not shared by the dinosaurs that laid the dinosauroid spherulitic type of egg. Such research in the field of dinosaur eggshell structure remains in its infancy, and further investigations within this relatively new field may provide significant new insights on the evolutionary history of dinosaurs.

Returning to our discovery, the eggshell in the Auca Mahuevo eggs is rather thin, somewhat less than one-tenth of an inch thick. Although this may seem thick in relation to a chicken's egg, it is indeed much thinner than the eggshell of other similar eggs that have been found in Patagonia and Europe. We were lucky in that the microscopic structure of the eggs we collected was preserved in perfect detail. By cutting thin slices through the shell and studying the shell structure with a stereo electron microscope, we could see that our eggs belonged to the spherulitic type, which suggested that they did not belong to theropods. This left sauropods and ornithischians as possible candidates. The Auca Mahuevo eggs were of a shape, size, and structure consistent with an egg type referred to as megaloolithid. A variety of megaloolithid eggs have been found in various late Cretaceous sites in South America, as well as in India and Europe. Much older megaloolithid eggshell has also been found in the late Jurassic of North America and Europe. Megaloolithid eggs have often been thought to belong to sauropod dinosaurs, but no such egg had ever been found with an embryo inside.

Based on evidence from sites found in many different parts of the world, we know dinosaurs laid their eggs in a variety of different patterns. Some dinosaurs laid their eggs in well-defined nests. Large concentric circles with pairs of eggs are typically found in nests of theropod dinosaurs, such as troodontids and oviraptorids, two kinds of maniraptors. The troodontids lay their eggs vertically, half-buried in the substrate, whereas oviraptorids lay them horizontally facing the center. Other dinosaurs laid their eggs in a spiral pattern within the nest. Still others laid their eggs in rather poorly defined patterns within a nest, and some laid them randomly across an area without a nest. These two latter patterns are typically the way in which megaloolithid eggs are found. In some Patagonian sites, megaloolithid eggs have been found in close clusters of six to ten eggs, whereas at a site in southern France, megaloolithid eggs were laid in what appears to be large semicircular arcs that do not seem to represent what we typically think of as a nest.

The distribution of the eggs at Auca Mahuevo was intriguing. In some places, the eggs appeared to be scattered over large areas, almost like a carpet of eggs. In other areas, they appeared to be more clustered, suggesting that nests may have been present. Clarifying the pattern of their distribution would clearly involve a lot of work and time. We were so busy just trying to collect egg fragments on the surface that contained fossilized patches of skin and quarrying more complete specimens of eggs that appeared to contain embryonic bones that we decided that studying the distribution of the eggs would have to wait until another season. Back in the laboratory, our top priority was to prepare the embryonic bones and see if they came from an ornithischian or a sauropod.

CHAPTER SIX

Finding Evidence in the Eggs at Our Forensic Labs

Dinosaur Embryos and Embryonic Skin—
the Rarest of All Dinosaur Fossils

Adult skeletons of dinosaurs are not that uncommon because their massive bones and durable enamel teeth stand a reasonable chance of becoming fossilized. Finding the fragile bones and skin of unhatched dinosaurs is much more rare, however, because their poorly calcified bones and delicate skin decompose rapidly after the embryo dies. At the time of our discovery, in fact, fossilized embryos were known from only a handful of different dinosaur species, despite hundreds of species having been discovered. The embryos of three kinds of meat-eating dinosaurs had been excavated from Cretaceous rocks of Mongolia, China, and Montana, including specimens of an oviraptorid, a therizinosaurid, and a troodontid, respectively. Less complete embryos of an unidentified species of meat-eating dinosaur had been collected from late Jurassic rocks of Portugal, and embryos from one kind of plant-eating duckbill, *Hypacrosaurus*, were known from Montana. But absolutely no fossils of embryonic dinosaur skin had ever been discovered before 1997. This poor representation of embryonic dinosaur remains explains why we were anxious to get our eggs prepared and see what was inside.

Fossil preparators are essential in any paleontological team. Even with the advent of modern techniques such as CAT scans, which allow paleontologists to peer inside fossil bones and skulls without manually

As with other hatchling dinosaurs, the skull of our embryos had a shorter snout and larger orbits than the adults. The reconstructed embryonic skull (bottom) is compared to what the skull of adult titanosaurs (top) may have looked like.

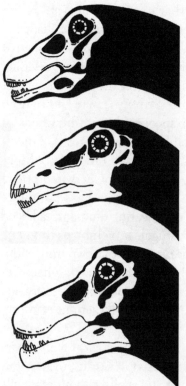

A reconstruction of the skull of a titanosaur (top) is compared to those of other sauropods with pencil-like teeth, such as *Diplodocus* (middle) and *Nemegtosaurus* (bottom).

cleaning away the surrounding rock, most fossils still require some manual preparation before they can be studied. Usually the surrounding rock must be removed by hand with small tools, such as dental picks, needles, miniature sandblasters, and miniature vibrating tools, although sometimes the rock surrounding a fossil can be etched away in a bath of dilute acid. Preparing delicate fossils such as our embryos requires enormous skill and patience. Under a microscope, the fossil preparator must slowly and carefully pick away the rock surrounding the fragile bones with the mechanical tools of the trade, then protect the bones by applying thin coats of transparent glue. To prepare just one tiny fossil embryo can take days or even weeks of painstaking work, but this had to be done to reveal the clues to identify the kind of dinosaur that had laid the eggs.

In late December 1997 and early January 1998, Marilyn Fox, our preparator at Yale University, made an important discovery. Inside one of the eggs, she uncovered some minute skull bones and miniature teeth. We hoped that the shape of the skull bones and the teeth would give us clues to identify the victims in the eggs, but this is never easy. Because the animals were so young, the bony tissue in their delicate bones was not well developed, thus obscuring comparisons with adult sauropods, whose bones are fully formed. Furthermore, the embryonic bones were crushed against the lower eggshell as the fluid in the egg leaked out and the bones became fossilized. Unfortunately, this would make identifying these small fossils difficult.

Luis took the train up to Yale to examine what Marilyn had uncovered. He brought several specimens back to New York just before Rodolfo came up from Argentina in the spring to see the fossils and help us write the scientific paper to announce our discoveries. One of our eggs contained a nearly complete skull, an important clue for identifying the kind of dinosaur that had laid the eggs because the skull bones of different kinds of dinosaurs are usually quite distinctive. The skull bones in our embryo were similar in shape to those of sauropod dinosaurs, the long-necked giants that include diplodocids, dicraeosaurids, barchiosaurids, and titanosaurs. This tiny skull was also remarkable because few adult skulls of sauropods have been discovered, let alone skulls of embryos. In fact, only a handful of adult sauropod skulls had been found in South America. Several bones in the skull of our embryo were not preserved well enough to be identified,

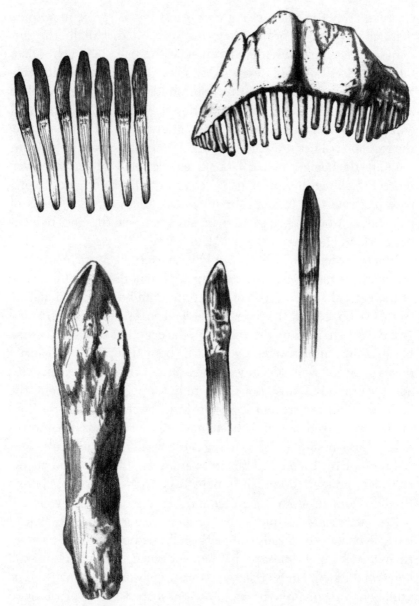

Only two groups of sauropods have pencil-like teeth: *Diplodocus* and its kin (top two images) and titanosaurs (bottom right two images). Other sauropods have thicker and more spoonlike teeth (bottom left).

but those that were allowed us to make a fairly accurate reconstruction of what the skull had looked like.

As in any other baby, the skull of our sauropod embryo was large in proportion to the size of the body, even though the whole head was only about two inches long. Likewise, the eye socket in our embryos was probably slightly larger in relation to the rest of the skull than the eye socket in adult sauropods, another characteristic of most infant animals. In addition, a couple of embryos had tiny, pencil-shaped teeth. The crown, or upper surface, was formed by enamel, the same extremely durable material that forms the crowns of human teeth, as well as those of many other animals. Dinosaur teeth come in a variety of shapes: some are designed like steak knives to cut flesh; others form tightly packed assemblages that serve as a grinding surface for macerating tough leaves and other kinds of vegetation; others are less specialized and shaped like tiny leaves. Despite all this dental diversity, however, only two groups of dinosaurs have pencil-shaped teeth—like the teeth from our embryos—and both are sauropods. Pencil-shaped teeth evolved once in the common ancestor of dicraeosaurids and diplodocids, but similar pencil-shaped teeth also evolved within titanosaurs. Which kind of sauropod did our teeth represent?

Titanosaurs are especially difficult to place on the evolutionary tree of sauropods. Some paleontologists argue that they are most closely related to diplodocids and dicraeosaurids, believing that the pencil-shaped teeth that are typical of all of these dinosaurs evolved only once. However, most students of sauropods argue that titanosaurs are most closely related to brachiosaurids. These researchers suggest that titanosaurs and brachiosaurs inherited a small claw on the first finger from their common ancestor, and that this claw was completely lost in some later members of the group. Additional characteristics in the hip and hind limb of these dinosaurs support the idea that brachiosaurids and titanosaurs are closely related. According to this argument, pencil-shaped teeth evolved twice—once in the common ancestor of dicraeosaurids and diplodocids and again in the common ancestor of titanosaurs. The fact that certain sauropods thought to be primitive titanosaurs—animals whose skeletons are very much like those of typical titanosaurs—lack pencil-like teeth supports this latter interpretation of a double evolutionary origin for this peculiar type of sauropod dentition.

The embryos lived at a time in which titanosaurs were common, especially in South America. As mentioned earlier, we had found their skeletons weathering out of the rocky cliffs near Doña Dora's *puesto*. The fact that titanosaurs lived around Auca Mahuevo at the same time as the embryos, however, does not constitute adequate evidence to conclude that the embryos were titanosaurs.

Our embryos had pencil-shaped teeth, but could we find unequivocal evidence in the bones and teeth of the embryos that linked them exclusively to either titanosaurs or to diplodocids and dicraeosaurids? Most experts believe that diplodocids and dicraeosaurids died out long before the late Cretaceous, but there is no consensus on this point. Central to this debate is a late Cretaceous sauropod from Mongolia called *Nemegtosaurus*, whom some regard as a titanosaur and others regard as a survivor of the diplodocid-dicraeosaurid group that survived long after most of its relatives had gone extinct. When the teeth of the upper and lower jaws met as *Nemegtosaurus* chewed, the crowns of the teeth were abraded such that nearly vertical wear surfaces formed, as in at least some titanosaurs. Surprisingly, a few teeth of our embryos had similar wear surfaces on their crowns. Even though the embryos could not have been chewing food before they hatched, they were obviously grinding their teeth in the same way they would have done when they ate after they hatched. Perhaps they were just exercising their jaw muscles to prepare for life in the world outside. Thus far, we had established that the embryos, *Nemegtosaurus*, and pencil-like-toothed titanosaurs all shared the same kind of dental wear surfaces. So if *Nemegtosaurus* was a titanosaur, the dental evidence would suggest that the embryos were titanosaurs. However, the only known fossil of *Nemegtosaurus* is a skull, which is not enough of the skeleton to determine if *Nemeg-tosaurus* belongs with the titanosaurs or with the diplodocids and dicraeosaurids, because little is known about the shape of the bones in a titanosaur skull. Unfortunately, therefore, we had no basis for comparison. There was at least one late Cretaceous titanosaurid roaming the ancient Gobi named *Quesitosaurus*, but no skull of this animal has yet been found. So we do not know what its teeth looked like or whether the whole skeleton of this creature was like that of *Nemegtosaurus*. Thus, although we can say that the embryonic teeth from Auca Mahuevo belonged to sauropods, it is not clear whether the embryos were titanosaurids, diplodocids, or another of their close relatives.

Another of our eggs contained several leg bones that fit up against one another. Although they did not prove helpful in precisely identifying our embryos, they established the approximate size of the embryo inside the egg. The thighbone, or femur, was about four inches long, twice as long as the skull of the other embryo, indicating that this embryo would have been ten to twelve inches long when it hatched. In adult sauropods, such as *Diplodocus*, the femur is four or five times longer than the skull. Although we cannot be sure whether the embryos were closer relatives of titanosaurs or diplodocids, all of these dinosaurs were enormous, and it is clear that our embryos would have grown into some of the largest animals ever to walk on earth. But what about the suspected fossils of embryonic dinosaur skin? Did those show similarities to previously known fossils of adult sauropod skin?

Fossils of adult sauropod skin have been known since the mid-1800s. The pattern of skin ornamentation in our embryos looked similar to that of other sauropod dinosaurs, such as *Diplodocus*. The skin in sauropods is formed by polygonal tubercles of varying size that do not overlap with one another. Recent discoveries have shown that these late Jurassic *Diplodocus* had a row of narrow spines running along their tails, like those of crocodiles, and some scientists argue that this series of spines would also have extended along the back and neck. None of the patches of fossilized skin that our crew found indicate the presence of spines on our embryos, but we believe that the triple row of larger scales found on our babies did extend along the entire tail, back, and neck. The difference between this pattern of embryonic scales and the pattern of spines in adult *Diplodocus* and its kin might suggest that the spines were used as a visual signal that allowed individual *Diplodocus* to recognize one another, perhaps at the time of selecting a mate.

The skin of our embryos exhibited a diverse array of scale patterns. In one, as mentioned above, a triple row of larger scales crossed a field of smaller scales. In others, we found scales arranged in rosette patterns, in which a circle of eight smaller scales surrounded a large central scale. Still other specimens revealed several triangular scales that converged toward a central point, like the petals of a flower. But

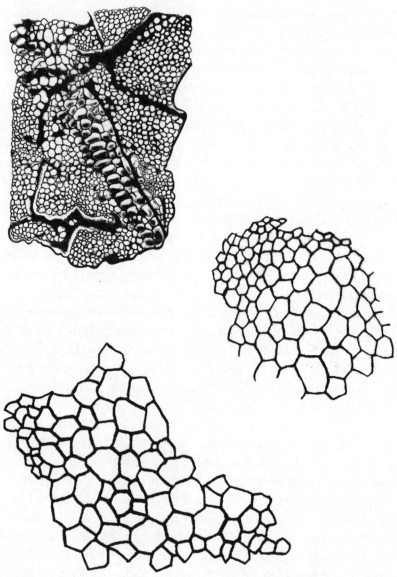

The pattern of scales of the embryonic sauropods (top left) resembled the arrangement of scales of the adults (center and bottom left).

unfortunately, it is not possible to say exactly where these scale arrangements were located on the body because the patches of fossilized skin did not overlap identifiable bones in the skeleton.

Several titanosaur specimens, all apparently close relatives of the late Cretaceous titanosaur from northwestern Argentina called *Saltasaurus*, were known from the same late Cretaceous rocks that we were exploring near Auca Mahuida. *Saltasaurus* is well known because it is covered with a fully armored skin, presumably for protection from the large meat-eating dinosaurs that lived at the time. Although armor is common in other groups of dinosaurs, including the stegosaurs and the ankylosaurs, it is not usually preserved with the skeletons of sauropods. In fact, it was not until the 1970s, when *Saltasaurus* was first found, that paleontologists realized that some sauropods possessed a covering of bony armor. We noticed that the pattern of armor plating in the skin of *Saltasaurus* was remarkably similar to the pattern of bumps on the skin of the embryos from Auca Mahuevo, which made us wonder whether we would find bony tissue within the skin of our embryos.

To study the skin of our embryos more closely, we wanted to take photographs using a stereo electron microscope and cut cross sections through the skin to study under other microscopes. Our SEM at the American Museum of Natural History required the specimens to be coated with a thin layer of gold or platinum paint before images could be taken, but we did not want to coat our real specimens. (Ironically, they are much more valuable to us without a gold or platinum coating than with one.) So we produced a rubber mold of our most complete patch of skin and made a resin cast of the skin from it. This perfectly replicated patch of embryonic dinosaur skin was coated with a thin layer of gold and then photographed. Luckily, we also obtained images of our embryonic skin outside the American Museum of Natural History using a more sophisticated electron microscope, which did not require the specimens to be coated.

Even at high levels of magnification, our embryonic dinosaur skin looks quite similar to the skin of other reptiles, like a blanket of round, scalelike knobs of similar size. In contrast to the scales on most modern lizards and snakes, the scales of our embryos did not overlap one another, just as in the case of fossilized skin from adult dinosaurs. In this respect, the skin of dinosaurs, including that of our embryos,

Foot-long baby sauropods hatched from a nest laid 80 million years ago in a remote corner of Patagonia, Argentina.

looks more like the knobby skin of Gila monsters than that of typical lizards. Folds in the skin on our embryos indicated that the skin was not closely attached to the muscles and bones. The folds probably formed in the joint areas between bones, just as skin folds at joints in modern animals.

As we said earlier, the scale patterns on our embryos were similar to the clusters of bony plates called scutes that had been discovered around the skeleton of *Saltasaurus*. The armor of *Saltasaurus* is formed by hundreds of small, closely packed, bony scutes—roughly the size of our fingernails—which are occasionally separated by four-inch-long, oval scutes adorned with a central ridge. This combination of large and small scutes occasionally forms roselike patterns. Both kinds of scutes are thought to have "floated" in the dinosaur's hide, although on certain areas of the body the scutes are so tightly packed that they would have formed a pavement of armor. Since the discovery of *Saltasaurus*, diverse scutes of other titanosaurs have been found in other parts of South America, Madagascar, and Europe. These discoveries have prompted the reinterpretation of scutes from earlier discoveries of titanosaurids, which had led to the belief that other kinds of armored dinosaurs such as ankylosaurids lived in South America.

Cross sections of the embryonic skin patches did not reveal bone. Nonetheless, the striking resemblance between the patterns on our embryos' skin and on *Saltasaurus* made us think that the bony armor of the adult titanosaur could have been formed as a one-to-one replication of the embryonic pattern of scales. This one-to-one replication is typical of modern armored reptiles, such as crocodiles and Gila monsters. Once again, our discovery suggested that the processes that control the development of modern animals were at work during the growth cycle of ancient dinosaurs. Although we believe that the scales of our embryos might constitute the model over which armor formed in adults, the smooth surface of our embryos' scales did not show any of the central crests seen in the larger scutes of *Saltasaurus* and other titanosaurs. It may be that these crests represent overgrowths developed during the formation of the bony scutes, or that perhaps crests were not present on the scutes of all sauropods.

All in all, our forensic studies revealed that the embryonic bones, teeth, and skin from Auca Mahuevo belonged to a large group of

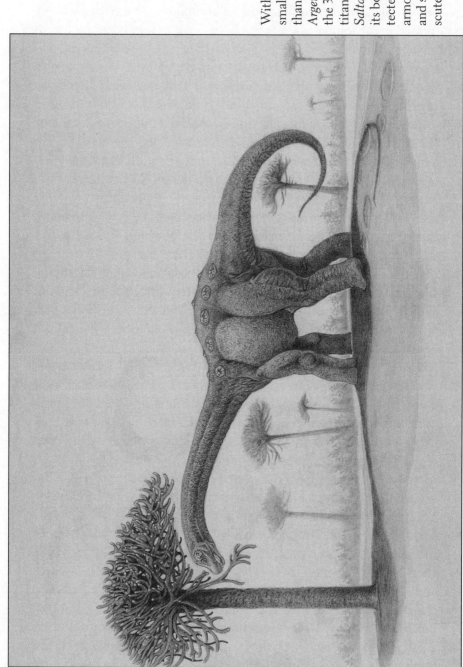

With a much smaller body than its relative *Argentinosaurus*, the 30-foot-long titanosaur *Saltasaurus* had its body protected by an armor of large and small bony scutes.

dinosaurs called neosauropods, which includes famous dinosaurs such as *Diplodocus, Camarasaurus, Brachiosaurus, Titanosaurus*, and *Argentinosaurus*. In a sense, we had discovered the smallest fossils of the largest dinosaurs. Our trip had succeeded beyond our wildest dreams, and our crew's discoveries represented several firsts for paleontologists. Most important, our discovery represented the first indisputable embryos of sauropod dinosaurs. Even though thousands of eggs attributed to sauropods had previously been found in France, Spain, India, Argentina, China, and other parts of the world, no *definitive* embryos of sauropod dinosaurs had ever been found until our team's discovery. A few small fossils from young sauropods had previously been discovered, and some paleontologists had argued that they represented embryos; however, they were either too big to really be embryos, or they were not found inside an egg—the definitive proof that a specimen had not yet hatched. At last, we could be certain that at least some sauropods laid eggs and that the large eggs previously identified as belonging to sauropods actually were sauropod eggs. Our embryos were also the first dinosaur embryos ever discovered in the Southern Hemisphere. Most dinosaur nesting grounds are concentrated in the northern continents, so the discovery of a large, new nesting ground in South America contributed important new insights to our knowledge about the reproductive biology of dinosaurs from the southern continents. Finally, the eggs also contained the first embryonic dinosaur skin ever discovered. So, for the first time, we could sense what it would have felt like to touch an unhatched baby dinosaur that would have grown up to become one of the largest animals ever to walk the earth. The discovery of embryonic sauropod skin also led to other important inferences concerning the biology of dinosaurs. As noted earlier, some scientists have argued that *Tyrannosaurus* was covered with feathers as a juvenile and that it lost those feathers as it grew into adulthood. This inference is based on the close evolutionary relationship between birds and *Tyrannosaurus* on the family tree of theropods. Sauropods are not theropods but are included with theropods in the larger group called saurischians. That adult sauropods and theropods more primitive than *Tyrannosaurus* and its coelurosaur relatives lacked feathers when adults was already known from several specimens with preserved portions of skin. Our discovery made clear that not all saurischians were covered with feathers during the early stages of their development, because our embryos were not.

Still, a lot more research needed to be done so that we could write and publish a scientific paper announcing our discoveries. As the preparation of the specimens continued throughout February, March, and April of 1997, we began to compile this research. Unlike articles in newspapers and magazines, scientific papers often take months or years to get published. The staff of the journal sent our paper would send it to other paleontologists for their comments and criticisms, a process called peer review, which helps authors improve the quality of their manuscript. Once other scientists make their

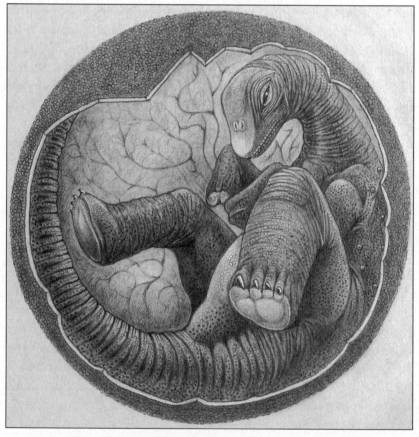

Our team discovered the first remains of the sauropod embryos inside eggs laid 80 million years ago.

comments, the editors at the journal decide whether the paper is important and accurate enough to publish. Once a paper is accepted, it can take several months to well over a year before the journal can produce and publish the paper. But before submitting our paper, we needed to try to figure out when and how the embryos had died.

CHAPTER SEVEN

Establishing the Time of Death

Evidence from Clocks in the Rocks

Just as fossils provide important information about life-forms that inhabited our world in the distant past, the rocks that they are preserved in constitute our only evidence to interpret when these ancient animals lived. To recount this part of our investigation, we need to flash back from the lab to the field.

What originally attracted us to explore the site that contained the eggs and embryos in Patagonia was the stunning visual beauty of the area's rocky outcrops. Layer upon layer of crimson sandstone and mudstone form a fantasyland of banded ridges and flats at Auca Mahuevo. This maze of ridges and ravines took nature millions of years to construct. First, the layers of sand and mud were laid down across the landscape when the dinosaurs lived. Then, these layers were buried under a thick blanket of subsequently deposited layers and remained under the surface of the earth for millions of years before the powerful tectonic and volcanic forces that created the mountains and valleys of the region lifted them back toward the surface. As they rose, rain and wind eroded the overlying layers of rock, leaving the ancient layers exposed on the surface once again. And once they became exposed, the rain and wind sculpted them into the breathtaking landscape spread out before us.

We knew from our experience in other expeditions that these kinds of exposures sometimes contained buried caches of fossil treasure. So as we drove down toward the badlands under the radiant morning sun that greeted the second day of our expedition, we could hardly wait to scramble out of our vehicles and begin prospecting. But once we had discovered the fossils, we also knew that these breathtaking outcrops also contained the key evidence for interpreting when the embryos died.

It is difficult for human beings to appreciate how long ago the animals preserved as fossils at Auca Mahuevo lived. The average life span of a person living in the United States is about seventy-five years. The United States itself is slightly more than two hundred years old. The earliest human civilizations based on agriculture from which we have discovered artifacts existed about ten thousand years ago. The Ice Ages ended about twelve thousand years ago, when such animals as saber-toothed cats, mammoths, mastodons, glyptodonts, and giant ground sloths went extinct. The earliest members of our human species lived around one hundred thousand years ago, and our earliest human relatives first walked the earth about 4.5 million years ago. All the dinosaurs, excluding birds, died out 65 million years ago. But the animals living at Auca Mahuevo lived millions of years before that. Based on previous studies of fossil animals that had been collected from the same layers of rock that were exposed in other parts of Patagonia, we knew that the rocks at Auca Mahuevo were deposited sometime between 70 million and 90 million years ago, a nearly incomprehensible span of time.

Geologists and paleontologists have developed a special time scale to serve as a geologic and evolutionary calendar. Based on major changes in the kinds of fossil organisms that lived at different times during the history of the earth, this time scale is divided into four major eras. From oldest to youngest, they are called the Precambrian (4.5 billion to 570 million years ago), the Paleozoic (570 million to 250 million years ago), the Mesozoic (250 million to 65 million years ago), and the era we now live in, the Cenozoic (65 million years ago to the present).

The earth first formed slightly over 4.5 billion years ago at the start of the Precambrian era, as the planets consolidated from rings of star dust orbiting our sun. The earliest fossils of ancient life that we have found were simple, single-celled organisms related to modern bluegreen algae, which lived in Precambrian oceans about 3.8 billion years ago.

The first of our vertebrate relatives did not appear until almost 3.3

billion years later, early in the Paleozoic era, and even then, these fish-like creatures, with segmented support structures resembling a back-bone, were not terribly imposing. They were only a couple of inches long, devoid of jaws, and probably fed inconspicuously on creatures that either lived in the mud at the bottom of the oceans or floated closer to the surface.

The first known animals and plants to move out of the oceans and live on land arose about 400 million years ago, during the Paleozoic era. Fossils of these animals and plants were preserved in rocks deposited in lush coal swamps, such as the rocks that now form the Appalachian Mountains in the eastern United States. Ancient relatives of horsetails, ferns, and tree ferns dominated the flora of these swamps, along with extinct groups of trees that have no close living relatives. Some of the early animals that colonized the land included different kinds of insects and other arthropods, who followed the plants in their invasion of the land. Dragonflies with wingspans of more than three feet patrolled the skies, and cockroaches over a foot long scavenged in the underbrush.

The earliest vertebrates to walk on land did not evolve until about 350 million years ago. Although often much larger, these amphibious creatures were built somewhat like salamanders and had to return to the water to lay their soft, membranous eggs. These first tetrapods—so named because they have four limbs—evolved from ancient relatives of the coelacanth, a lobe-finned fish that was long considered to be extinct until it was discovered living in the waters of the Indian Ocean in the 1930s. Early reptiles and relatives of mammals appeared on the scene about 300 million years ago, near the end of the Paleozoic era, but the origin of dinosaurs still lay millions of years in the future, after the start of the Mesozoic era.

The Mesozoic era, sometimes called the Age of Large Dinosaurs, is divided into three different periods, the earliest of which is the Triassic period, which lasted from 250 million years ago until about 206 million years ago. As we've already mentioned, the earliest known dinosaurs lived in the Triassic, about 230 million years ago, when most of the continental masses were fused into a huge single continent called Pangaea. The earliest known mammals originated near the end of the Triassic.

Next comes the Jurassic period, when the supercontinent of Pan-

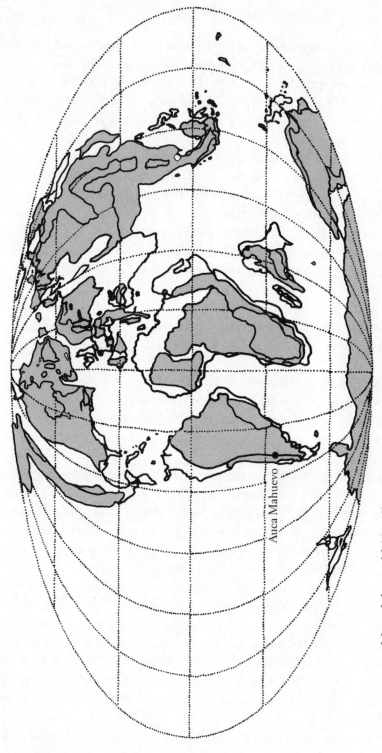

Map of the world 80 million years ago, when sauropod dinosaurs laid their eggs at Auca Mahuevo. Areas in gray illustrate emerged continents.

Auca Mahuevo

gaea began to split apart, which lasted from 206 million years ago to 144 million years ago. This period saw the evolution of most of the largest dinosaurs that ever lived. We have already introduced some of them, including the sauropods, a group of enormous herbivores such as *Apatosaurus* (formerly called *Brontosaurus*), *Diplodocus*, and *Brachiosaurus*, which thrived in the tropical and subtropical climates that predominantly characterized this interval. But this period also witnessed the evolution of such terrifying carnivorous forms as the twenty-foot-long *Allosaurus* with its three-foot-long skull and four-inch-long, serrated teeth. In addition to these large dinosaurs, the earliest-known flying descendants of dinosaurs—birds—began to compete for dominance in the skies. Their primary rivals were pterodactyls and other flying reptiles, whose bodies were not covered with feathers and whose wings were constructed very differently from those of birds.

But the dinosaurs at Auca Mahuevo lived during the last period of the Mesozoic era, the Cretaceous period, which lasted from 144 million years ago until 65 million years ago. During the Cretaceous, rifting between the subcontinents of ancient Pangaea gave rise to the modern continents that we recognize today. This period also witnessed the origin of our modern biota, for many groups of living vertebrates arose then, as well as the flowering plants and their ubiquitous insect pollinators. In North America, this period saw the evolution of the duck-billed dinosaurs such as *Anatotitan*, the horned dinosaurs, such as *Triceratops*, and the fearsome carnivore *Tyrannosaurus rex*. *Tyrannosaurus* was long recognized to be the "king" of dinosaurs, as denoted by its species name. At almost thirty-five feet long, its slender but powerful hind legs probably made it a relatively swift and agile predator for its impressive size, although some paleontologists have recently argued that it primarily filled the ecological role of scavenger. Regardless of the feeding niche it filled, *Tyrannosaurus*'s four-foot-long skull, studded with eight-inch-long teeth shaped like steak knives, made it the most imposing carnivore on the North American continent. In Argentina, as mentioned earlier, the Cretaceous saw the evolution of the largest dinosaur yet discovered, the herbivorous *Argentinosaurus*, and a ferocious carnivore that may well have been larger than *Tyrannosaurus*, *Giganotosaurus*. Even as we write this book, fossils are being excavated near Plaza Huincul in

Neuquén province from gigantic meat-eaters that are even larger than *Giganotosaurus*.

We knew that the animals at Auca Mahuevo lived between 70 million and 90 million years ago, near the end of the Cretaceous. But 20 million years is a long time, and we wanted to pin down the time of the embryos' death more precisely. This required that we collect small samples from many of the rock layers at the site. We knew that the layers had been laid down one on top of the other so that the lower ones were older than the higher ones. This fundamental geologic principle, called superposition, established several centuries ago by early students of earth history, allows one to establish the relative age of fossils in a sequence of rock layers. It was, therefore, critical for us to record which layer or layers contained the fossils at Auca Mahuevo, because fossils from lower layers were older than fossils from higher layers. The question was, exactly how old were they?

Figuring out the exact age of the fossils from Auca Mahuevo was difficult and involved detailed scientific analysis. First we compared fossil animals at Auca Mahuevo with fossil animals from other localities where the age was known. If the kinds of animals were very similar, then the fossils from Auca Mahuevo were assumed to be about the same age. Using just the dinosaurs, however, it was hard to tell, because no other rock layers at other sites contained many of the same kinds of dinosaurs. But some fossils in rock layers higher in the sequence near our site were of animals that had lived in the ocean, such as clams, snails, and microscopic plankton. These suggested that the dinosaurs at Auca Mahuevo were at least 70 million years old. In addition, rocks that underlie the layers at Auca Mahuevo contain dinosaurs that were estimated to be older than 90 million years. How was the age of these animals established?

Some rocks containing similar fossils in other parts of the world also contain ancient layers of weathered volcanic ash, and the volcanic ash contains small crystals of minerals that were formed during the volcanic eruption. Some of these crystals contain atoms, including uranium and potassium, that break apart into other atoms at a constant rate in a process called radioactive decay. The atoms that break apart are called parent atoms, and the atoms that result are called daughter atoms. Using sophisticated scientific instruments, geologists can measure how long it takes for half of the parent atoms to break up into their daughter atoms; this amount of time is called the half-life. Geologists can also measure how many parent and daughter atoms are present in the small crystals that were formed when the volcano erupted and can use these data to calculate the age of the layer of volcanic ash and estimate how old fossils in nearby rock layers are. Sometimes rock layers that can be dated by this method of radioactive decay do not actually contain fossils. Nonetheless, if the fossils from other rock layers in the stratigraphic section are above the dated layer, we know that they are younger, and if the fossils are below the dated layer, we know that they are older.

These methods of radiometric age estimation are based on the same process of radioactive decay utilized in the carbon-14 technique for dating ancient human remains and artifacts. The carbon-14 technique is not applicable for dating rocks and fossils older than about

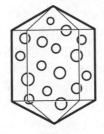

Composition of crystal at time of formation.

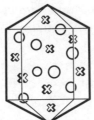

After one half-life, half of parent atoms have decayed to daughter atoms.

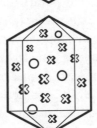

After two half-lives, half of remaining parent atoms have decayed to daughter atoms.

Diagram showing basic concept of radiometric dating.

o=parent atoms in crystal lattice
x=daughter atoms in crystal lattice

fifty thousand years, however, because then the half-life of carbon-14 (C-14) is only about 5,700 years. In objects older than about fifty thousand years, there is usually not enough C-14 remaining to obtain accurate measurements for calculating the age. The half-lives of the uranium (U-238) and potassium (K-40) are much longer. Half of the U-238 atoms decay to lead (Pb-206) atoms in about 4.5 billion years, and half of the K-40 atoms decay to argon (Ar-40) atoms in about 1.3 billion years, which makes these techniques suitable for estimating the age of much older objects but less useful for dating more recent objects.

We hoped that the greenish sands at Auca Mahuevo might contain minerals that were erupted out of volcanoes when the greenish sands were formed, so we collected many samples for analysis back in the laboratory. Unfortunately, none of the rock samples that we brought back from these greenish layers surrounding the fossil eggs contained mineral crystals that had been erupted out of volcanoes. Instead, they represented tiny pieces from a large pool of molten rock that had cooled and crystallized deep underground, so we could not be sure that these crystals formed at the same time as the rock layers that contained our fossils. Hence, we could not estimate the age of the rock layers and eggs through radioactive decay. It was frustrating to fail in this initial attempt, but that is the nature of scientific research. We knew that we would have to expand our search for layers of ancient volcanic ash into other adjacent regions.

In addition to collecting rock samples for possible dating through radioactivity, we had collected rock samples for dating through magnetic analysis.

The earth is like a giant bar magnet with a north pole and a south pole. Throughout the known history of the earth, our planet's magnetic poles have commonly reversed directions, such that the magnetic end of a compass needle that points north today would have pointed south then. Recent geologic research has suggested that such reversals of the magnetic poles can happen over a period of a few hundred years. This may seem like a long time to humans, but it is only a brief instant in comparison to the vast expanse of geologic time. In the last 65 million years, for example, the magnetic poles have switched positions about thirty times, and the last time the poles switched was about 750,000 years ago. But how do geologists know this?

Some kinds of sedimentary and volcanic rocks contain minute particles made of iron-bearing minerals, such as magnetite. When these particles settle out through a water column or cool within a body of magma, they align themselves with the earth's existing magnetic field. So, in rocks, these microscopic, magnetic mineral grains reflect the direction of the magnetic field at the time that the rocks were formed, which we can then analyze in a magnetometer.

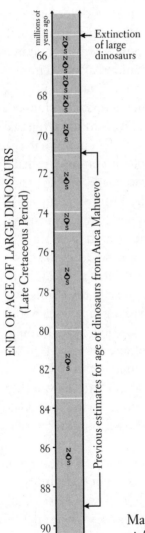

Magnetic polarity indicated by the rocks exposed at Auca Mahuevo.

Why the earth's magnetic field occasionally reverses direction is not well understood, but the sequence of these reversals has been recorded in the earth's rocks. By combining these sequences of reversals with ages from rock layers that can be dated through radioactive decay, geologists have reconstructed a calendar of when the magnetic poles were oriented as they are today and when they were reversed. Using samples of rock from Auca Mahuevo, from the layers containing the eggs and embryos, we might be able to estimate the age of the rock by comparing our data to this global magnetic calendar.

We had carefully collected rock samples from eight different layers associated with the fossil eggs. First, using a small pick and a rock hammer, we had dug into the hillside to expose a two-foot-square area of unweathered rock. (Weathered rock could easily yield unreliable magnetic information.) Next, we used a hand rasp to plane a horizontal surface between three and five inches across on a chunk of rock. Fortunately, the mudstone and siltstone layers in the sequence were fairly soft, so the rasp worked effectively. To check that the planed surface was level, we used a special geologic compass, called a Brunton compass, which has a leveling bubble inside. Once a truly level surface had been planed off, we used the Brunton compass to mark an arrow pointing toward the north magnetic pole, thus showing how the chunk of rock was oriented in the ground in relation to the earth's present magnetic field. Once the arrow was marked, we had to carefully dig the chunk out of the ground without breaking it. To keep it in one piece during the trip back to the lab, the chunk was wrapped in aluminum foil and secured with masking tape. Finally, the sample was labeled with a number, and the position of the sample was recorded on the drawing of the stratigraphic section, so that we would know where it had come from once the magnetic analysis was conducted. The magnetic analyses, which might determine whether the rocks had been formed during a "normal" or "reversed" period in the earth's magnetic history, had to wait until we got the samples back to the lab.

The rock samples that we collected for magnetic analysis were somewhat more helpful in establishing the time of the embryos' death than were the samples collected for radiometric analysis. Carl Swisher and Gary Scott of the Berkeley Geochronology Center determined that the rocks containing the sauropod embryos at Auca

Mahuevo were deposited during an interval of time when the earth's magnetic field was reversed. These results helped to narrow down the 70-to-90-million-year range based on marine fossils preserved in higher layers of rock and dinosaurs from lower layers of rock. The presence of rocks at Auca Mahuevo that had formed during an interval when the earth's magnetic poles were reversed means that the rocks had to have been formed less than 83 million years ago.

We knew this because the global magnetic calendar documents that the poles were oriented as they are today between about 100 million years ago and 83 million years ago. Because the Auca Mahuevo rocks were formed when the magnetic poles were reversed, they had to have been formed after the end of that long interval. In addition, recent studies of fossil pollen found in rock layers just above the egg-bearing layers at Auca Mahuevo suggest that the pollen is between 76 and 81 million years old. Although we have yet to sample for pollen in these same layers at Auca Mahuevo, the same sequence of rusty mudstone and sandstone layers that contains the eggs is present where the pollen was found, about one hundred miles to the south of our site. Since these pollen fossils were higher in the rock sequence than the egg-bearing layers, the pollen had to be younger than the eggs and embryos. The magnetic poles of the earth were not reversed between 76 and 79 million years ago, but they were reversed between 79 and 83 million years ago. We concluded, therefore, that the eggs and embryos from Auca Mahuevo were probably between 79 and 83 million years old.

The 4-million-year span in our age estimate might still seem like an eternity. After all, almost the entire history of our human lineage is contained in the last 4 million years of geologic time. Most fossil sites from these more recent periods of earth history can be dated more precisely because the analytical uncertainties in estimating the age are not as great. Sites containing remains of large extinct dinosaurs are at least 65 million years old, however, so, given the analytical uncertainties, we would have to be satisfied that the time of death had been refined a bit through our magnetic analyses. But another large mystery remained to be solved: What was the cause of death?

CHAPTER EIGHT

Establishing the Cause of Death

Uncovering Clues in the Rocks

To figure out what had killed all the unhatched sauropods in the eggs, we again had to appeal to the rocks that contained the fossils. In addition to providing the only direct evidence to estimate the time of the embryos' death, the picturesque layers of rocks at Auca Mahuevo preserved key clues for interpreting the cause of their death, which appeared to be directly related to the environment in which the sauropods had laid their eggs. The evidence for interpreting what environment the dinosaurs had lived in would be gleaned from observing the different kinds of rocks that formed the layers at Auca Mahuevo.

Throughout the world today, rocks are either being eroded or formed on the surface of the earth in an ongoing cycle that has continued since the early days of earth history when the first rocks were formed. In areas that are relatively high and steep, including the mountains and hills that dominate all the continents, rocks are weathered and eroded by rain, ice, and wind. Chemical reactions promoted by molecules dissolved in rainwater help break the bonds between mineral crystals forming rocks near the surface. Aided by the destructive activity of plant roots, as well as the expansion and contraction of ice as it forms and thaws, loosened debris is eroded by runoff from rains and the force of the winds. In colder regions, glaciers can also serve as powerful agents of erosion, scouring away great quantities of mountainous terrain. In areas that are relatively low and depressed, including river valleys and lakes, some of the material eroded away from higher regions is deposited in layers on land and in

lakes by rivers and streams, as well as by winds and the melting of glaciers. Other material is washed or blown off the continents to be
deposited in layers at the bottom of the oceans.

In all of these modern settings, different types of debris, called sediments, are laid down under different environmental conditions by
erosion and deposition. We can observe how rain and ice erode
mountains and carry the residue down from the heights in rivers
and streams to be deposited across floodplains and within ocean
basins. We can also document the kinds of sediments that are being
deposited on the floodplains and in the oceans. By comparing sedimentary layers forming today to the kinds of rocks we find in ancient
rock sequences, we can identify the kind of environment in which the
ancient sediments were deposited. This is possible because sedimentary rocks are simply ancient sediments that, with time and the
action of several geological processes, have become petrified. The sediments that we see being transported and deposited today will eventually become the sedimentary rocks of tomorrow.

Such observations led some early students of geology to a rather
startling, yet intuitive, conclusion about how the earth has evolved
since it formed. In essence, they reasoned that the present is the key
to the past, because the processes of erosion and deposition that we
can observe operating today have acted more or less similarly throughout the vast expanse of geologic history. This concept is a fundamental
dictum of geology called the principle of uniformitarianism.

Illustrative examples are provided by the many volcanoes that
have erupted in the past across the globe, generating both lava and
ash. Once cooled, some lava forms a kind of volcanic rock called
basalt that has a characteristic texture and composition of minerals.
When ancient sequences of rocks contain this same type of hardened
lava, we know that it was erupted out of an ancient volcano, whether
or not the volcano still exists.

In 1980, for example, a tremendous eruption occurred in southern
Washington State at Mount St. Helens. Millions of tons of volcanic
ash were blasted several miles up into the atmosphere, where winds
carried the ash to be deposited across the states of Washington,
Idaho, Montana, and even into the Dakotas. The ash formed a layer
of powdery dust as it settled out of the air on distant cities and landscapes. Again, its texture and mineral composition is quite distinc-

tive, and similar layers can be found in ancient sequences of rocks, testifying to their origin during volcanic eruptions. When petrified, these layers of ancient volcanic ash are often called tuffs. We had hoped to find these tuffs in the sequence at Auca Mahuevo in order to provide environmental clues and to establish the age of the rocks through radioactive decay analyses.

Similarly, rivers and streams can be seen to deposit layers of gravel, sand, silt, and mud as they make their journey from steep mountain canyons across gently sloping floodplains. The kind of sediment that a stream can carry depends on the velocity and turbulence of its currents. Swift, turbulent currents flowing down steep slopes can transport boulders, as well as smaller particles of sand, silt, and clay, but less turbulent currents cannot carry such large sedimentary debris. Consequently, coarse, heavy sediment is often deposited by swift rivers and streams near the mouth of a steep mountain canyon, where the slope of the stream flattens out. These boulders, pebbles, and coarse sand grains are dumped into the valleys adjacent to the mountains as alluvial fans, while much of the fine sand and mud is carried on downstream. Immense alluvial fans can be seen today at the mouths of canyons leading into Death Valley in California. As mentioned earlier, the gravel-rich layers behind Doña Dora's *puesto* represent the kind of coarse gravel that was deposited on alluvial fans adjacent to some ancient hills or mountains on the Cretaceous Patagonian landscape when titanosaurs roamed across southern South America.

On more gently sloping plains farther away from the highlands, often on the coastal plain, slower, less turbulent rivers and streams deposit lighter sediment in stream channels, flood basins, and lakes. The coarser sand often forms sandbars in the stream channels. Finer silt and clay is often deposited outside the banks of the channels during floods, when the rivers and streams overflow their banks and the slower currents carry the silt and mud far away from the main channels before it settles out. Such deposits of sandbars in channels and mud on the floodplain away from the channels have been observed throughout many modern river systems, and similar layers of sand and mud can be found in many ancient rock sequences. These are prime places to look for fossils of dinosaurs, as well as other animals and plants that inhabited the ancient floodplains, because their bones and leaves could quickly be buried by flood debris before they could decay or be scavenged. In fact, many of the well-known dinosaurs from North America and other parts of the world, including *Tyrannosaurus* and *Triceratops*, come from sedimentary rocks formed in these kinds of environments.

Lakes also form in depressions on the floodplain, and they tend to accumulate their own distinctive layers of sediment. Away from where streams feed into the lake, thin layers of muddy clay, which floated far out into the lake before settling down through the water, are formed on the bottom. These thin layers, called laminations, often form finely striped rocks once the sediment is compacted, and this kind of rock is found in many ancient sequences of rocks. This environment also provides excellent conditions for the preservation of not only fish and other animals that live in the lake, but also vegetation and carcasses of dead animals that float out into the lake. The fine mud that settles out of the water can bury these remains quickly and preserve skeletons in which almost every bone remains in a natural, lifelike position. Often, the water at the bottom of a lake is depleted in oxygen, which reduces the number of bacteria and scavengers that destroy sunken carcasses, and in these instances, soft tissues such as hair, muscles, and internal organs stand a better chance of being preserved. Ancient lake sediments have produced some of the most exquisitely preserved fossil specimens of dinosaurs, insects, plants, and

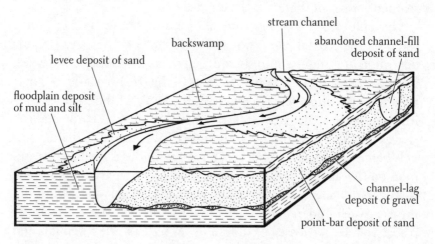

Diagram showing features in floodplain environment.

other organisms that have ever been found. Famous fossil sites such as the Cretaceous Liaoning lake sediments in northeastern China and the Tertiary Green River lake beds in Wyoming represent good examples of these ancient, fossil-rich environments, and similar rocks would yield critical clues about what dinosaurs preyed on the sauropods at Auca Mahuevo.

Finally, after the river flows into the sea, any remaining sand and silt are deposited near the shoreline, but finer-grained silt and mud can be carried farther out into the ocean basin. Out in deep ocean waters, the shells of microscopic, single-celled plankton settle to the bottom after the organisms die to form layers of limy mud on the bottom that are eventually compacted into limestone.

Given this brief introduction, let's once again look at the rocks containing the sauropod eggs and embryos at Auca Mahuevo. The eggs and embryos were entombed in rusty brown layers of silt and mud, which were mixed in with coarser layers of greenish and reddish brown sandstone. The alternation of layers containing sandstone and mudstone closely resembled the kinds of sediments that are deposited across floodplains by streams and small rivers. Thus, evidence from the rocks indicated that between 79 million and 83 million years ago, the dinosaurs at Auca Mahuevo lived on a broad, gently sloping floodplain, crisscrossed by shallow streams and rivers. This floodplain formed as South America drifted away from Africa due to the enormous forces generated deep within the earth as the result of plate tectonics, the geological process that drives the continents across our planet's surface. Thin layers of sandstone were the geologic clues that documented the ancient presence of shallow stream channels and their sandbars. The sand was not too coarse, and few pebbles were present, which suggests that the streams were not as swift and turbulent as the ones that deposited the coarse gravel found in the ridges behind Doña Dora's *puesto*. Clearly, there were no large hills or mountains nearby to provide steep stream gradients and large pebbles or boulders. In addition, the thickest layers of sandstone were about three to four feet thick, suggesting that the streams were not terribly deep. Over time, these streams migrated back and forth across the floodplain, cutting and then filling in the channels with sandbars, but the eggs were not found in these layers of sandstone.

Fossils of the eggs and embryos were found in the finer-grained layers of mud and silt. What might this clue tell us about the cause of death? These layers represented silt and clay particles that were carried over the banks of the streams when they flooded and inundated the adjacent lowlands. Similar events are common in our modern world; every year, heavy rains in many parts of the globe unleash torrential floods that inundate major river valleys, often destroying populous cities and leaving a blanket of mud wherever the flood passes. Objects as large as cars and houses can be transported or buried by the onslaught. Although the streams on the primeval Patagonian floodplain were not large enough to generate such enormous floods, they nonetheless overflowed their banks from time to time, carrying a blanket of mud and silt to be deposited as currents slowed, away from the main channels. Eighty million years ago, no cities stood in the way of these floods—only the vegetation that grew on the plain and occasionally a nesting ground of sauropods.

It appears that the dinosaurs intentionally looked for places away from the streams in safer areas of the floodplain to lay their eggs. Eggshell is actually pretty durable, and fragments can be carried over substantial distances without being completely destroyed. However, we didn't find any fragments of eggshell in the sandstone formed from the bars in the channels that were active at the time of the nesting, and the eggs did not appear to have been broken or even transported by the floodwaters. So, the currents depositing the mud could not have been too strong, and the eggs in the nesting ground could not have been very close to active channels. Why did the dinosaurs apparently avoid the areas near the active stream channels? Perhaps it was just chance, but it may also have been because they had some sense that the streams could destroy their nests. In any case, at times this strategy, if there was one, didn't work. For the floods, although not very powerful, nonetheless brought a wave of sudden death to the nesting ground.

Since the fossil eggs and embryos that we found in 1997 were exclusively buried in the layers of mudstone, not the sandstone, we concluded that the eggs and embryos had quickly been buried when the streams flooded. (Years later, we would find eggs that had been laid in sand rather than mud. But the sand formed the bed of a long-abandoned channel at the time that the dinosaurs laid their eggs, and the eggs had later been buried by mud during a flood.)

When the floods buried the eggs in their debris, it not only began the fossilization that preserved the eggs but also either drowned or suffocated the helpless embryos inside under a massive sheet of water or a sticky blanket of mud. Thus the floods not only killed the embryos, but by burying the eggs in a layer of protective mud, prevented them from being exposed to plunder by scavengers, which would have greatly decreased their chances of being preserved as fossils. This might sound pretty awful, but for us paleontologists living 80 million years later, it was a stroke of good luck.

At last, the probable cause of death had been established, and we were now ready to submit our scientific paper to a journal for publication. Because our discoveries represented so much new information and so many firsts for paleontology, we decided to try to get our paper published in one of the most prestigious scientific journals in the world, *Nature*. We sent them our paper in the middle of 1998, but it would be several more months before the paper was actually published. Although we thought that the media might be interested in our discoveries when the paper finally came out, we were not prepared for the overwhelming reaction.

CHAPTER NINE

Our Fifteen Minutes of Fame

Is Auca Mahuevo the Real Jurassic Park?

In November of 1998 just before Thanksgiving, two articles about the nesting site at Auca Mahuevo announced what we had found. One was the scientific paper in *Nature*; the other was a popular article that was published in *National Geographic*. The National Geographic Society had helped to fund our 1997 expedition, and they wanted to let the public know what we had found.

To let other media outlets know about the discovery, we also scheduled a news conference at the American Museum of Natural History on the day that the articles were released. The museum had sponsored the expedition, and at the time we both worked there. We were fortunate that Rodolfo was able to come up from Argentina to join us for the news conference.

Setting up a news conference requires a lot of work. The public relations department of the museum spent several weeks contacting journalists and correspondents at newspapers, magazines, television stations, and radio stations, letting them know that we intended to announce a major discovery about dinosaurs. Most of this responsibility fell to Elizabeth Chapman, the director of media relations for the museum at the time.

Elizabeth is a diminutive dynamo of energy and enthusiasm for anything scientific that goes on at the museum. Her job was to encourage media coverage of the institution's scientific research, and there is no subject that she loves more than dinosaurs. She developed that love as a small child. When she used to visit her grandparents in Pittsburgh, they would take her to the Carnegie Museum, and she would demand that her grandmother read every word on every label in the dinosaur halls. One year, her grandmother tired of this exercise and failed to read the last few sentences on one of the labels. Elizabeth immediately protested that her grandmother had not read all of the text. Elizabeth had memorized all of each passage.

Elizabeth had joined us for a few days at the end of our expedition in November 1997 so she could get a better idea of what we had found. She had wandered around the site with childlike amazement and glee, picking up eggs off the surface of the flats and inspecting them for remains of fossilized skin and embryos. From her curiosity and wonderment, we knew that this was one of the most satisfying moments of her career.

When Elizabeth had the chance to help us announce our discovery, she pulled out all the stops. For weeks before the press conference, she contacted media outlets all over the world to let them know that a significant announcement was in the works. The day before the conference, she faxed hundreds of press releases to reporters around the world. Of course, there was no guarantee that all the correspondents would be interested, so she was always cautious with us about how many news organizations would respond.

On the day of the press conference, we arrived early to set out some of the fossils for the reporters to see and then calmly went back to our offices to wait. We were sure that some reporters and television crews would show up, but we didn't think it would take too long. As the hour arrived, we walked into the room where the conference was to be conducted. To our amazement, the room was packed solid with cameras and reporters. At least fifty reporters and ten television cameras were crammed throughout the room, with camera technicians jostling each other for prime positions as we walked in to begin our presentation. We glanced at each other in stunned amazement.

Elizabeth began to brief us about who had shown up; the list was staggering. The *New York Times, Time, Newsweek,* and the Associated Press had all sent reporters, as had numerous other local New York newspapers. CNN, ABC, NBC, CBS, the Associated Press, the Discovery Channel, and several local New York television stations had sent camera crews. Even several publications and media outlets from

South America attended. The blinding lights from the cameras flashed on, and the show began.

The juxtaposition of the desolate scenes from the field and the media frenzy in the conference room seemed rather surreal. Little did we know when we had traveled to that remote corner of Patagonia that we would bring back fossils of the tiniest giant dinosaurs, who would become instant media celebrities 80 million years after they had died.

After we were introduced, we showed some slides and a video of the site as we described the kinds of fossils we had found. This presentation took about half an hour before we opened the floor up for questions. The questions and subsequent interviews went on for another three hours. By midafternoon we were fried and retreated for lunch to our offices. But the wave of publicity was just beginning to crest. During the news conference, dozens of reporters who had not been able to attend called the museum to ask questions and get quotes from us. The phone calls continued throughout the afternoon. By evening, we were exhausted and hoarse, but we still had to ride down to the BBC studio to do a live interview for their late-night news on radio and TV. By the time that was over, we had been on stage for almost nine hours straight. It was definitely time for a drink.

That evening, numerous members of the field crew and the media, along with some of our other colleagues and friends, gathered at a Spanish restaurant downtown. InfoQuest, one of the foundations that had sponsored the expedition, had arranged for an informal party to celebrate our announcement. When we arrived at the restaurant, at about 8 P.M., several guests said that they had already seen clips from the news conference on TV. It was all quite gratifying, but the almost instantaneous nature of the coverage was also rather numbing. We hoped that, in all the commotion, we hadn't said anything too stupid in front of the cameras. About 10 P.M., the party broke up, and we went home. But the onslaught of coverage would continue the next day: at five-thirty the next morning, we had to show up bright and cheery for an appearance on *Good Morning America*.

We awoke to a torrent of publicity. In addition to the scientific report that was published in the November 19, 1998, issue of *Nature*, the *National Geographic* article appeared. The story made headlines around the world, with front-page articles in *The New York Times* and the *Los Angeles Times*; television segments on ABC TV's *Good Morning America*, the *CBS Evening News*, and BBC TV; articles in *Time* and *Newsweek*; and dozens of stories in newspapers and magazines, on radio shows across North America, Europe, Japan, and Latin America. In all, Elizabeth estimated that this tsunami of media attention inundated about 100 million people.

All the attention was great, but some of the coverage was completely unanticipated. A few days after our announcement, for example, *The Daily Show,* a parody of the nightly television news on the Comedy Central network, surprised us all with their interpretation of our discovery. They suggested that what we had really found was the world's oldest known abortion clinic. Another one of our favorite accounts appeared in one of the tabloids. The paper had printed pictures of our eggs and the embryonic dinosaur skin and quoted a fictitious paleontologist, who claimed he had discovered living embryos of the giant sauropods, which he was now incubating. Eventually, when they hatched, he said, he intended to set up a game preserve where they could grow up and reproduce—a real Jurassic Park. We wish the report had been accurate, but it was clearly an exercise in science fiction rather than science. Nonetheless, similar questions arose during many of our interviews, which was not surprising given all the recent advances in cloning and the popularity of *Jurassic Park*—both the novel and the movie.

Let us take a moment to address this issue here. In the movie *Jurassic Park*, numerous kinds of extinct dinosaurs are brought back to life by reactivating the genes of these dinosaurs after the genes have been recovered from the bodies of mosquitoes or other bloodsucking insects preserved in amber. The scientists extract the dinosaur blood from the fossilized insects, separate the dinosaur DNA, which contains the genes, from the blood, and use that to re-create the dinosaurs. But how realistic is this scenario in light of science's recent success in cloning sheep, mice, and even monkeys?

It is undeniably true that many kinds of small fossil animals and plants can be preserved in amber, which is simply sap from ancient trees that has been buried in the earth for millions of years and has hardened into a yellowish, clear solid. Insects are one of the most common kinds of animals to be found in amber, because they became trapped in the sticky sap while foraging for food. Amber is an unusually good material in which to find fossils because delicate structures are often preserved, such as fine hairs on the insect's body and the pattern of veins in the wings. These soft tissues are not commonly preserved in fossils buried in sandstone or mudstone. The exceptionally complete fossilization of insects in amber provides scientists with many extraordinary clues to identify ancient insects and to analyze where their groups fit on the evolutionary tree.

In fact, however, no Mesozoic insects in amber have ever been found with blood inside them, and most insects in amber lived much later than the dinosaurs seen in the movie *Jurassic Park*. As mentioned earlier, dinosaurs originated in the Triassic period. The earliest-known dinosaurs, such as the small carnivores *Herrerasaurus* and *Eoraptor* from Argentina, lived about 230 million years ago. Unfortunately, we have never found amber with insects inside from rocks that are this old. The next period during the Age of Dinosaurs was the Jurassic, when the first giant, plant-eating sauropods, such as *Brachiosaurus*, lived, as well as fierce carnivores such as *Allosaurus*. But no Jurassic amber with biting insects inside has ever been discovered. So despite the name of the book and the movie, there is not even any insect-bearing amber available to use in trying to clone dinosaurs from the Jurassic period.

There are, however, some biting insects preserved in Cretaceous amber. This was the age in which *Tyrannosaurus*, *Velociraptor*, *Triceratops*, and *Ornithomimus* lived, as well as the sauropods from Auca Mahuevo. A few small biting insects called midges are known from amber this old, but none have been found with blood in them, and none of these have been found at Auca Mahuevo. At present, the oldest known biting insect preserved in amber is about 125 million years old. It is possible that paleontologists will eventually find Triassic, Jurassic, and Cretaceous insects with blood inside them, but it would be rare and, based on experience, unlikely.

One reason for this improbability is that blood breaks down quickly after an animal dies. Rotting actually begins within minutes after death, which is why bodies must be embalmed if a person is to be buried and is also why blood that is donated for operations must be maintained under strict temperature controls. So even if a blood-filled insect were to be discovered in amber, the blood would have had to have been perfectly preserved in amber for more than 65 million years to be useful for the purpose depicted in the movie *Jurassic Park*. This would be unlikely because, as the amber is buried deep within the earth during fossilization, temperatures can easily reach several hundred degrees, which would surely destroy the blood cells. An additional problem is that the amber itself often contains natural cracks, increasing the chances that contaminants in the groundwater that flows through the rocks in which the amber is buried will damage the dinosaur genes in the blood.

Another problem is that the insect would have had to die before the dinosaur's blood was digested in its stomach, because digestion would also have damaged the dinosaur's genes. What's more, there is no guarantee that the insect's last meal would have been the blood of an extinct dinosaur, given that many other kinds of animals also lived during the Mesozoic, including turtles, lizards, crocodiles, pterosaurs, birds, and even our own early mammalian relatives. Furthermore, it would be no easy task to remove the dinosaur's blood and genes from the insect inside the amber. For the purposes of cloning in the movie, the dinosaur's genes would have to be kept separate from the genes in the insect's body. But since we would have to cut through the insect to get at the dinosaur's genes, it would be difficult to keep the tissues of the two animals separate and uncontaminated.

But assuming that all this could be done, molecular biologists have recently developed a technique called polymerase chain reaction (PCR) to duplicate the genes enough times to produce enough genetic material with which to do some experiments. Most of the research done in labs around the world using this technique is directed toward comparing the genetic codes of different animals in order to establish their evolutionary relationships. To date, the oldest insect DNA that has successfully been duplicated is genetic material from a termite preserved in amber between 25 and 40 million years ago. That was a remarkable accomplishment achieved by molecular biologists working at the American Museum of Natural History in New York. Nonetheless, that termite lived long after the Mesozoic ended 65 million years ago.

The problems don't end here, either. An animal's genetic code is like a book containing information about how to build that particular animal. The chromosomes are like the chapters of the book; genes are sentences; and nucleic acids are letters of the alphabet. Because

DNA molecules break apart so easily, the best that could be hoped for in separating dinosaur DNA from the insects is to find two or three hundred letters of the genetic code still stuck together. This would constitute an ordered sequence representing less than a millionth of the whole genetic code—not nearly enough to make any sense of the whole book. So finding a piece of extinct dinosaur DNA inside the body of an insect preserved in amber would be like finding a sentence from a long book that had been cut up into millions of pieces. With just that one sentence, we would have almost no chance of understanding the meaning of the book, because we would be missing millions of other sentences, thousands of other paragraphs, and dozens of other chapters. The scientists in *Jurassic Park* avoid this problem by adding DNA from living animals to the dinosaur DNA. However, we do not know *nearly* enough about how DNA works to be able to patch pieces of living and extinct DNA together like this. Since each kind of animal has its own genetic code, the result would almost certainly be gibberish, not a sensible book specifying how to resurrect an extinct animal.

But let's make believe that we were somehow able to get all the pieces of the genetic code for an extinct dinosaur out of the fossilized insect in amber. Could we re-create the dinosaur then? Again, the problem would be one of putting the letters, sentences, and chapters back together in the correct order to make a sensible book. Thus, even if we had all the sentences (all the short segments of DNA in the genetic code), our challenge would be similar to putting a huge puzzle back together when all of the pieces are similarly shaped. It would be virtually impossible to put them all back together in the correct order. But even if all the DNA segments could be assembled in the proper order, the genetic code is only one aspect of what is needed to re-create an extinct dinosaur or any other living animal. The mother's body provides a whole chemical and physical environment that is needed for the embryo to develop within before birth. The scientists who cloned Dolly the sheep, for example, used the body of a living female sheep to nourish the developing embryo. But we have no mother's egg within which an extinct dinosaur embryo can develop. Thus, we are missing an essential component in the process. Without the supporting environment inside the egg, we would end up with just a bunch of chemicals floating around in a test tube.

This problem was fancifully solved in *Jurassic Park* by placing the dinosaur DNA inside an egg cell of a female crocodile. This makes a bit of evolutionary sense because the crocodile is the closest living relative of extinct dinosaurs, except for birds. It would have made even more sense to stick the dinosaur DNA inside the egg cell of a female ostrich, but even this wouldn't have worked. Not enough is known about what triggers and controls an embryo's growth and birth to successfully duplicate an animal that has been extinct for 65 million years.

So our chances of re-creating an extinct dinosaur are not good. In fact, there is realistically no chance at all. Could that change in the future? Perhaps . . . in the far distant future, but many momentous scientific breakthroughs would have to be made to do so. Realistically, those breakthroughs are impossible for us to envision and, in all probability, will never occur. Molecular biologists and paleontologists are constantly asked whether extinct dinosaurs can be cloned. As our press conference demonstrated, reporters often like to probe endlessly, trying to force us to admit that the possibility exists. After extensive badgering, when one of the museum's scientists was asked whether he thought there was just a slight chance of this all eventually coming to pass, he thought for a second and exasperatedly replied, "Yeah, and someday monkeys might fly out of my butt, too." Naturally, we hope, for our colleague's sake, that the chances of that occurrence are, indeed, equally low—even in the years to come. Although many scientists would have offered a more politically correct answer, his response reflected the overwhelming consensus among scientists that an actual Jurassic Park is unlikely to become a reality.

Having survived the media blitz, our thoughts once again turned to the field. Our initial investigations had raised many other biologic and geologic questions that we would have to solve to compile a complete picture of the events preceding and following the catastrophe, and we wouldn't be able to find those answers with the evidence we had in the lab.

CHAPTER TEN

Our Return to the Scene of the Catastrophe

The 1999 Expedition

Just as modern investigators must often return to the scene of an accident to gather more evidence after their initial investigation, we also needed to reexamine the site of our discovery. There were still many mysteries about how the sauropods had laid their eggs: Did they lay them in discrete nests or scatter them randomly across the surface of the floodplain? If they were in nests, how many were laid at one time? Did all the eggs belong to one sauropod species, or did multiple species use the same site? Did they return to the nesting site year after year or use the site only once?

To help solve these mysteries, we needed some special expertise, so two specialists on dinosaur eggs came with us. One was Frankie Jackson from Montana State University and the Museum of the Rockies in Bozeman. Before traveling to Patagonia, she had spent more than ten years collecting and studying dinosaur eggs, embryos, and nests in Montana for Jack Horner, whose team of collectors was responsible for completely reinvigorating the study of dinosaur eggs and embryos in the late 1970s with their discovery of Egg Mountain. Frankie had served as Jack's chief collector and preparator throughout much of the collecting and research at the site. A tall, thin, self-effacing woman, Frankie speaks with a slow Southern drawl, which she honed to perfection in her native Alabama. Her passion for and knowledge of dinosaur eggs is boundless.

Our other egg specialist was Gerald Grellet-Tinner from the University of Texas at Austin. Gerald was a student of another of our close colleagues and friends, Timothy Rowe, a professor in the Department of Geosciences. Tim's research over the last fifteen years has shed considerable new light on the evolution of both primitive dinosaurs called ceratosaurs and the early relatives of our own group of vertebrates, the mammals. His pioneering studies of vertebrate fossils through the use of CAT scans have also made a significant contribution to the field of vertebrate paleontology. Gerald's thesis had taken the study of bird eggs in a totally new direction by applying the methods of cladistics to explore what the structure of bird eggshell can tell us about how birds evolved from one another. Initial results are promising and his work shows that the structure of eggshell contains important clues for understanding the genealogical relationships of egg-laying organisms. Gerald plans to extend his novel approach to the study of dinosaur eggshell while continuing his research at the Natural History Museum of Los Angeles County. His tall, sturdy stature, along with his gregarious personality and wild stories of past adventures, makes him a natural for fieldwork, and his keen sense of observation, as well as his geological background, proved invaluable to our efforts.

We were also joined by two intrepid correspondents, Malcolm Ritter of the Associated Press and Tom Hayden of *Newsweek* magazine, who helped bring our continuing efforts to the attention of the public.

Our return to Patagonia was sponsored by the Natural History Museum of Los Angeles County. Since our last expedition, Luis had become an associate curator in that museum's Department of Vertebrate Paleontology, and his museum played an essential role in arranging for two new vehicles from Honda through its Office of Corporate Sponsorship. The National Geographic Society, the InfoQuest Foundation, and the Ann and Gordon Getty Foundation once again generously provided essential funding for the trip. Finally, after more than a year of preparation, planning, and research, we were ready to return to the scene of the catastrophe. We planned to spend the entire month of March 1999 at the site, searching for clues that would help us solve more of the mysteries of the ancient calamity.

After flying to Buenos Aires in late February, we gathered equipment and supplies before leaving for Auca Mahuevo on the twenty-seventh. Our two-day trek to the site was largely uneventful, at least until we left the highway for the final seventy miles of dirt roads that wound deep into the desert. As we turned north for this last leg of the trip, we noticed a dark thunderhead looming just above the horizon. Within half an hour, the first sprinkles began to splatter on our windshields, but more ominously, the sky above us had transformed itself from a hot but tranquil blanket of blue into a roiling cauldron of dark, greenish gray tumult. Part of this storm had passed just ahead of us, and large puddles began appearing in the low spots of the dirt road. As we drove on, the puddles transformed into deep ponds, which presented no problems for the larger, four-wheel-drive vehicles, but our entourage included a small Fiat sedan owned by Luis's brother, Ezequiel Chiappe. He had willingly volunteered to help us for a week, and his physical strength was a great asset during the early days of heavy lifting and excavation. But his car was not built for these kinds of road hazards; with only six or seven inches of clearance, it struggled to ford the deep ponds. The rain was torrential, and in places the road was covered by almost two feet of water. While Lowell was filling in for Ezequiel during the last stretch to camp, the wipers on the Fiat stopped working, and he had to stick his head out the window to see where he was going. Water splashed over the hood as we plunged into the pools on the road, and the car could barely cross to the next dry patch. Finally, we crossed over a small ridge that led down into a wide ravine; at the bottom, we could see a flash flood crossing the road. We were stuck in the middle of nowhere.

Initially, we considered turning around and trying to outflank the storm, but a semi truck soon pulled up behind us, and the driver told us that the road we had come on was now washed out behind us. For the moment we were trapped: we had no other option but to wait until the torrent ahead of us, which turned out to be over two feet deep, receded. It seemed rather ironic to us that, 80 million years after floods had wreaked havoc on the sauropods' nesting ground at Auca Mahuevo, our own expedition had, at least temporarily, fallen prey to the same natural hazard on the parched desert landscape of modern Patagonia.

Within a couple of hours, the thunderstorm passed to the southeast, and the stream ebbed somewhat. With the larger vehicles leading the way, we ventured into the current. We had seen when the semi

moved through that the water was only about a foot deep, but to Ezequiel's dismay, the Fiat was pretty well spent. We basically had to tow
it through the receding water. But once we had forded the stream,
which was still about a quarter mile in width, the road improved dra-
matically.

By 7:30 P.M. we limped into the tiny, rustic village of Auca
Mahuida, where we intended to camp this year. Less than ten people
live in this small outpost, nestled in a picturesque ravine of reddish
sandstone layers. Like Doña Dora and her family, the inhabitants now
eke out a living off the land by raising sheep, goats, horses, and cattle;
however, the town initially developed around the activities at a nearby
asphalt mine. Once a relatively prosperous community of several
hundred miners, Auca Mahuida experienced a sudden eclipse in the
late 1940s when a fire killed seventeen workers and the authorities
decided to close the mine. The village was abandoned, and the modern community evokes a ghostly image of its former heyday. Most
houses have crumbled, and the desert has reclaimed most of the
surroundings. Nonetheless, the village is only a ten-minute drive
from our site, so our daily commute would be much shorter than the
thirty-minute trek from Doña Dora's. In addition, the people of the
village allowed us to use a partially crumbled and abandoned house.
A few rooms still had roofs intact to protect our equipment and
food, but we soon realized that each time we entered our sanctuary
would be an adventure in itself when Rodolfo was struck by a falling
brick, which opened an inch-long cut across his forehead. In spite of
this precarious shelter, the town was furnished with a working well, so
we would have fresh water without shipping it in from Neuquén or
Plaza Huincul.

Rodolfo and his crew greeted us as we drove into town. Within an
hour, we had pitched our tents and begun to settle in. Much to our
delight, Rodolfo's crew was preparing a celebratory *asado* to mark
our reunion in the field. As we dried out by the fire and toasted our
arrival, the tribulations of the afternoon deluge quickly gave way to
enthusiastic discussions of past triumphs and impending adventures. We could hardly wait to get started.

We set out for the site on the morning of March 1 and spent most
of the morning introducing our new crew members to the surroundings, including where we had found fossils the previous year and the

basic geological features. The paleontological and geological teams then split up and began their separate operations.

The first task for our fossil collectors was backbreaking. Luis and Rodolfo wanted to expose a large area of the egg-bearing layer at the quarry where we had found the embryos on our previous trip, but this would involve considerable excavation. At the nose of the ridge where the embryos had been found, most of the mudstone layer containing the eggs was buried under three to four feet of other rock. This spot was not easily accessible, so we could not use a bulldozer to clear off the overburden. We all knew that a more primitive approach would be required.

With everyone pitching in, we hauled the picks and shovels up the ridge and set to work. Our goal was to expose an area of the egg layer about thirty feet long by fifteen feet across, which would require us to remove about thirty to forty tons of overlying rock. It was a long, hard day, but with about ten people helping, we were nearing the top of the egg layer by the end of the afternoon. Once we got close to the top of the egg-bearing layer, we would trade our picks and shovels for smaller rock hammers, chisels, and dental tools to excavate the eggs without breaking them. But we could begin that job the next day.

A mile away on the other side of the field area, Lowell and the geological team were doing some digging of their own, though fossil eggs were not their quarry. Several hundred feet of sandstone and mudstone layers lay on top of the egg-bearing mudstone, but we had not had time to measure or collect magnetic samples from these layers during our first trip. Our intent now was to continue measuring the rock layers above the egg-bearing layer and collect more samples for magnetic analysis. Starting just above the same layer that contained the egg quarry, we began working our way up the ridges.

Lowell described the rock types and measured the thickness of the layers for the stratigraphic section. Julia Clarke and Alberto Garrido, an Argentine graduate student in geology from the University of Córdoba, collected rock samples for paleomagnetic analysis. Alberto is a handsome, soft-spoken young man in his late twenties who, in addition to his general geological training, is familiar with the rocks of this region because he grew up in the area and has volunteered as one of Rodolfo's field assistants since he was a teenager.

Every fifteen feet or so, we would look for a suitable layer of silt-stone or mudstone from which to collect a magnetic sample. Each sample required about thirty minutes to excavate, measure, and wrap in aluminum foil. By the end of the day, we had collected about eight magnetic samples and measured about two hundred feet of rock layers above the main egg layer. In the process, we had located a new egg layer about seventy-five feet above the mudstone layer where the first eggs and embryos had been found. Our project to measure these layers and collect more magnetic samples would continue into the next day, but for now, it was time to rejoin the paleontological team and return to camp.

Unfortunately, when we arrived, a surprise was awaiting us. Our provisions had been too much of a temptation for the dogs of the village, who had knocked open the door of the house where we had stored our food and helped themselves to most of the meat we had brought. Well-gnawed bones littered the yard outside the house. Obviously, even in this rather remote corner of the Patagonian desert, we were going to have to beef up the security system around our home to fend off these four-legged intruders.

On the morning of March 2, both teams picked up where they had left off the previous day. Under Luis and Rodolfo's direction, the fossil collectors began the more delicate operation of removing the thin layer of mudstone that still overlay the eggs in the quarry, and by the afternoon, the tops of eggs began appearing on the surface of the expanded area of excavation. It was reassuring to realize that all the hard work of the last two days had not been wasted.

Most of Frankie and Gerald's work on our project involved uncovering a large area of the egg-producing layer to see where the fossil eggs were preserved within the layer. If they were not randomly placed but formed in discrete clusters, it would suggest that the eggs had been laid in nests. In addition to Frankie and Gerald, two students were instrumental in helping with work in the quarry. Gareth Dyke was a graduate student at the University of Bristol in England. He volunteered in order to gain a kind of field experience unavailable to him at home, and his determination during the long hours of quarrying contributed greatly to the trip's success. Anwar Janoo was a postdoctoral fellow in the Department of Ornithology at the American Museum of Natural History. He hails from Mauritius

Island in the Indian Ocean east of Madagascar, which was once inhabited by the famous dodo bird—a large, flightless relative of pigeons that was exterminated during the early days of European exploration and colonization of the Indian Ocean. Anwar is one of the world's foremost experts on this fascinating bird, which represents one of the most compelling stories in modern science about humanity's role as an agent of extinction. Anwar served on our expedition as a keen-eyed collector, as well as an excellent cook.

An intriguing fact was that the eggs were appearing within an interval of the mudstone that was about two feet thick: some were near the top of that interval, whereas others were near the bottom. At this early stage of the excavation, we were not sure whether the eggs were distributed in just one thick layer or whether there were actually two layers of eggs closely packed on top of each other. Only several days of patient excavation and Frankie's detailed mapping of the eggs' distribution in the quarry could help us answer that question.

On the other side of the field area, Lowell, Julia, and Alberto continued measuring the rock layers and collecting samples for magnetic analysis. This work was rather tedious because, as mentioned earlier, all one can do in the field is collect the samples. Analyses to determine whether the rocks had formed when the earth's magnetic field was normal or reversed would have to wait until we returned to the United States and worked on them in the laboratory. But in the midst of this tedium, a punctuating moment of unexpected discovery emerged.

We had discovered the first eggs and embryonic skin on our second day in the field during the expedition in 1997. For some reason, our second day in the field at Auca Mahuevo on our 1999 trip also turned out to be magical. After a morning of measuring rock layers and collecting magnetic samples, the geological trio headed down a ridge to return to their vehicle to drive to the quarry where lunch was waiting. On the way down the ridge, Alberto walked past some light beige fragments of rock weathering out of the hillside. Bending down to pick up one of the larger fragments, he could immediately see that these weren't simply small rocks; they were fragments of fossil bone. After calling Lowell and Julia over to examine the fossils, the crew began to collect some of the larger chunks. Clearly, at least some of the fragments were from small vertebrae a few inches long, though none of them knew what kind of animal the fragments came from. Nonetheless, our geologists were excited because Auca Mahuevo, although rich in dinosaur eggs and embryos, had not yet yielded any skeletons of adult dinosaurs. Perhaps that would now change.

After putting a number of the larger chunks in a plastic bag, the crew drove back to the quarry. A large tarp had been strung between two of the vehicles to provide a bit of shade from the withering noonday sun. As we all began to eat, the bag of fossil fragments circulated among the paleontologists to see if anyone could identify the fragments. Immediately, Luis and Rodolfo became intrigued, and after a few moments of close inspection, they agreed on several basic points. First, the fragments represented tail vertebrae of a dinosaur. Second, the tail vertebrae did not seem to belong to a sauropod, but instead came from either a meat-eating theropod or a plant-eating ornithischian. This point was particularly exciting because with all the eggs of sauropods that we had found, it seemed most likely that any skeletal remains discovered would also be from sauropods. Third, we should examine the site more closely to see if more of the skeleton lay buried beneath the surface.

The next day, Luis, Rodolfo, and the geological team examined the remaining vertebrae fragments on the surface. Then we began clearing away the loose dirt around some of the larger fragments with a paintbrush to see if more bones lay buried under the surface. Within a few hours, our efforts had paid off. One by one, four more tail vertebrae slowly appeared. A tone of elation and anticipation permeated our conversation, but there was also some trepidation as it became clear that we would face a massive excavation effort if the whole skeleton lay beneath the dirt.

We were certain that, as new bones were exposed, we were progressing from the back of the tail toward the front. We could all see that these new tail vertebrae were arranged in a line that pointed into the small ridge on which we were working, and that each new one that we uncovered was larger than the previous one, as we followed the tail into the hill. At this point, we were sure that whatever remained of the skeleton was leading into the small hill that it sat on. In terms of preservation, this was excellent news: perhaps only the end of the tail had weathered away before Alberto had found the skeleton. But to find out, we would have to do some digging to remove the rock above where we hoped that the skeleton was buried.

Rodolfo and his crew volunteered to excavate the new dinosaur skeleton, and on March 4, they began by shoveling a foot or two of mudstone off the area. Then, as they carefully picked through the remaining mudstone with pocketknives and other small tools, more tail vertebrae appeared. By the end of the day, all the rest of the tail vertebrae had been exposed, along with the hipbones and some of the large bones of one hind leg. To our relief and great joy, the bones were all well preserved, and they fit up against one another in the same positions they had occupied when the animal had died.

With the back half of the animal now exposed, Rodolfo and Luis could see the shapes of the bones, providing some clues for identifying it. The bones of the tail and hind leg were almost identical to those found in *Carnotaurus*, one of the meat-eating theropods called abelisaurs. We were all thrilled because remains of abelisaurs are extremely rare, and many parts of their skeleton had never been found before. Rodolfo was particularly ecstatic: as a specialist on large predatory dinosaurs, he had been hunting for more than a decade for a complete skeleton. It looked as if his quest might finally be fulfilled.

At the quarry, work was also progressing nicely under the supervision of Frankie and Gerald. By the end of the day, sixty eggs had been exposed and mapped. Frankie had divided the surface of the quarry into one-meter squares and recorded the position of each egg in three dimensions from one corner of the quarry. To our knowledge, no other study had been able to document the distribution of so many eggs, and we hoped that our good fortune might allow us to make some important advances in understanding how sauropods had laid their eggs. With work at both quarries yielding spectacular results, it was time to celebrate: the sizzling of that evening's *asado* was accompanied by the popping of champagne bottles.

On the following day, Rodolfo and his team continued to probe through the mudstone at the dinosaur quarry, and they found even more bones from the skeleton. The arms were small in relation to the rest of the body, although they were proportionally slightly larger than the arms in *Carnotaurus*. Nonetheless, the shapes of the arm bones were very similar. Some vertebrae from the neck were also uncovered, which suggested that the skull might be preserved some-where under the surface. In all, it appeared that, except for the end of the tail, which had weathered away before Alberto had found the skeleton, all of the bones might be present.

While reflecting on our new discovery, we realized that another piece of the ancient puzzle of Auca Mahuevo had fallen into place. Although we had still not found a good skeleton of the adult sauropods that had laid the eggs, we now had a pretty good idea about who had probably served as their primary predator. The parts of the skeleton that had been exposed showed that this menacing meat-eater was about twenty feet long. Based on the long, powerful hind legs and short arms, it clearly walked exclusively on its two back legs and probably weighed between one and two tons. Although Alberto's spectacular discovery lay many feet above the egg layer, there was lit-tle doubt that the species to which it belonged inhabited the area when the Auca Mahuevo sauropods gathered to nest. It would have been a terrifying adversary. Such an animal could easily have taken down a young sauropod by itself, and if the predators pooled their efforts in packs, even adult sauropods could have been at risk.

Quarrying of the eggs also continued on March 5. By now more than eighty eggs had been exposed, and we hoped that we would soon have several hundred eggs excavated and mapped. Unfortunately, the fickle Patagonian weather had other ideas about how we would spend the next two days. By the end of the afternoon, clouds were building on the horizon, and we covered both quarries with large tarps just in case the torrid summer heat was temporarily interrupted by another torrent from the approaching fall.

During that night, the rains arrived, though not the kind of tor-rential downpour that we had experienced earlier on our way to the site. It was just a gentle, steady rain that lasted for two whole days. Frustratingly, however, the ground was so muddy that we could not get to either of our quarries. Even if we had made it, it would have been impossible to work because the wet clay of the egg-bearing mudstone would have stuck tenaciously to the tools and fossils, risking damage to the eggs if we had tried to excavate them. To deny the rain an opportunity to dampen our spirits, we headed to Plaza Huincul to see some new exhibits at Rodolfo's museum. Since the last time we had visited, the museum had finished a new, full-scale mount of the monstrous meat-eater *Giganotosaurus*, which was spectacular. This new exhibit, along with a quick shower at a local motel, worked wonders to refresh our strength and enthusiasm, so we headed back to camp with renewed vigor.

The trip turned out to be both long and a bit harrowing. Night had fallen and the headlights on one of our older vehicles, which Lowell was driving, began to fail. It became impossible to see more than a few car lengths in front of us, and the headlights of oncoming cars were completely blinding. We stopped to get the vehicle checked out and bought a new battery at the last gas station before we left the paved road for the long stretch of dirt roads leading into the desert, but the problems soon reappeared. Lowell and Sergio Saldivia had to drive the last seventy miles to camp slowly in order to see the dirt roads at all, so we didn't get back until about 2:00 A.M. Ironically, it turned out to be fortunate that the rain continued throughout that night and all the next day, because we had a chance to sleep in, do some laundry, take a lazy afternoon nap, and catch up on writing our field notes. By about noon on March 8, the rains finally ceased, and we were once again eager to go.

The rains had caused some minor damage in the quarries, but fortunately, the fossils were not affected. The pits were quickly drained so that work could continue. At the egg quarry, Luis, Frankie, Gerald, and the rest of the team focused on removing some of the rain-softened overburden so that the excavation could be expanded to the full thirty-feet-by-fifteen-feet that we had originally planned to map. In the abelisaur quarry, Rodolfo's crew continued to expose more of the skeleton.

Lowell, Julia, and Alberto, having finally finished measuring the rocky layers and collecting magnetic samples at the site, turned their attention to figuring out how many egg layers actually existed. Clearly, at least two separate layers were present: one contained most of the eggs on the flats and in the quarry, whereas the other lay about seventy-five feet higher in the sequence. But were there still others, as yet unrecognized? A closer examination of the eggs weathering out on the flats revealed that two separate layers of mudstone, separated by about five or six feet of unfossiliferous sandstone and mudstone, actually contained eggs. In addition, another isolated cluster of eggs was discovered about twenty-five feet below the layer that contained the quarry. This meant that at least four separate layers contained eggs, which was an important clue for us in our attempts to understand the nesting behavior of the gigantic sauropods, as we will discuss later in more detail. Gerald collected samples of the eggshell from each layer. By studying the patterns of ornamentation and chemical composition of the shell in the lab, he hoped to determine whether more than one kind of sauropod had laid the eggs in the different layers.

On March 9, some of the crew members took a long drive to Neuquén to buy supplies and take care of some bureaucratic tasks, but Rodolfo's crew worked on in the dinosaur quarry. More and more frustratingly, we had yet to discover the skull, or even any evidence that a well-preserved skull was present. The small bits and pieces of skull bones found up to this point suggested that the skull might have been badly damaged either before or during fossilization. But it was also possible that the skull had become detached from the neck and was buried somewhere close by. When Marilyn Fox arrived at the site, she assumed the job of digging carefully through the mudstone to find out, but despite several days of painstaking excavation, she could still not find a large piece of it.

On the tenth, a sense of tedium descended over our operation, mixed with a growing sense of anxiety. We were about halfway through the field season now, and our wealth of success began to seem like a mixed blessing. It seemed impossible to get both the eggs and the large new dinosaur skeleton collected and analyzed. The work at the egg quarry seemed interminable; there were eggs everywhere we dug. Our hopes for a fairly complete abelisaur skeleton had been realized, but to excavate such a large skeleton would require more than a week of intense digging and plastering by Rodolfo's crew. And once the plaster jackets had been constructed around the bones, the heavy jackets would have to be rolled over, plastered on the bottom, and lifted onto a vehicle that could transport them back to the museum in Plaza Huincul. Some sort of heavy equipment, such as a bulldozer or a crane, would be required for the lifting. But where would we find that in the middle of the desert? Rodolfo left camp for Plaza Huincul to see what kind of arrangements he could make.

To make matters worse, the recent rains had triggered a totally unexpected population explosion among the insects of the area. Clouds of mosquitoes began to hatch from the small ponds and pools that were left scattered across the desert. Our normally peaceful evenings of relaxed conversation quickly degenerated into long hours of swatting and slapping as we tried to fend off the buzzing, bloodsucking hordes. Our only advantage was that they slept during the day, which allowed our work to proceed without these annoying interruptions.

The next two days were spent wrapping up odds and ends, especially around the egg quarry. Frankie precisely documented the position of more than two hundred eggs using the system of grids laid out in the quarry. In addition, Luis, Gerald, and Gareth built plaster jackets around several clusters of eggs so that we could collect them for further study. One was huge, containing forty or fifty eggs and probably weighing almost a thousand pounds. Natalia Klaiselburd, who was released from her duties in the quarry to prospect, found three more fragments of eggs that contained patches of fossilized skin. Another embryo was found in an egg near the quarry. Throughout this period, the mosquitoes were still on the rampage.

With the eggs in the quarry now completely exposed, Lowell and his geological team spent most of the morning of the thirteenth examining the site. We focused our attention on a puzzling geological phenomenon: the mudstone in the egg layers was laced with smooth, shiny, grooved surfaces that looked similar to slickensides, the surfaces one sees in fault zones where blocks of rock slip past one another as movement occurs along the fault. However, the slickensides in the mudstone at Auca Mahuevo were not large and continuous, the way one would expect if they had been created by a large fault that extended for hundreds of yards or miles across the countryside. These were only a few inches to a foot long. Some of the magnetic samples from the mudstone that we had collected in 1997 were yielding screwy results, which we thought might have resulted from the movement of the mudstone blocks along the small slickensides. But we didn't have any idea what had caused the slickensides, since they weren't large enough to be major faults.

Fortunately, David Loope, a geology professor from the University of Nebraska, had arrived the previous evening, and we described what we had been seeing to him. Dave possesses a calm and considered demeanor, as well as an entertaining sense of humor that is as dry as the Patagonian desert, and he is a widely recognized expert on ancient sand dunes who had accompanied us on expeditions to the dinosaur-rich deposits of the Gobi Desert in Mongolia. There, his observations were critical in helping us understand how the beautifully preserved fossils had come to be killed and preserved by massive, water-soaked sand avalanches that had slid down the dune faces during occasionally heavy rainstorms. His extensive knowledge about how different kinds of rocks form would also prove invaluable to our studies of the rocks that entomb the eggs and embryos at Auca Mahuevo.

Dave immediately thought that the slickensides might be related to a geologic process called vertisols, which operates in some soils. He explained that, in clay-rich soils, alternating cycles of wetting and drying can lead to expansion and contraction of the material in the soil. The expansion occurs when it rains and the clay absorbs the moisture, which causes blocks of the soil to move upward, creating the slippage surfaces or slickensides that we had observed. This "vertical" movement occurs because the direction of least resistance is upward, since rock or dirt is present on the sides and below the soil, but only the atmosphere or a thin layer of soil overlies the expanding clay. The vertisols would probably have formed within decades or a couple of centuries after the floods deposited the mudstone.

The vertisols appeared to have some other important implications. First, they offered additional evidence that the area occasionally received substantial amounts of rain, alternating with substantial periods of dryness. Since we suspected that floods had killed the embryos, the vertisols would help establish that rains potentially generating floods did occur. Also, vertisols could easily have led to the development of a hummocky surface on the floodplain, with small bumps and depressions littering the landscape, and such depressions might have made good places for the sauropods to lay their eggs. Finally, the vertisols could explain why many of the eggs were somewhat squashed and some of the egg clusters appeared to be mixed up with one another. Movement along the slippage surfaces could have created these effects by crushing the eggs trapped in the moving blocks and displacing others.

Frankie's mapping in the quarry gave us a good idea about how individual eggs were distributed in clusters, but the quarry was too small to document how the clusters were distributed across the land surface. To determine how the clusters were distributed, Luis,

Frankie, and Gerald went back to the flats. There, we selected a 120-foot-by-90-foot rectangular area and marked all the enclosed egg clusters with colored balloons, which proved slightly difficult because, although some of the clutches were easy to identify, others had almost completely weathered away. To make sure that a clutch was actually present, we decided to mark only clusters that contained remnants of eggshells arranged in a circle and imbedded vertically in the ground. Using a tape measure, we divided the selected area into squares that were thirty feet on a side, thereby creating a gigantic grid. Then, we once again plotted the precise position of each of the egg clusters within the squares. When the grid was constructed and the clutches marked, the green and blue balloons created a rather surreal scene across the naked landscape of the Patagonian desert, as if we had created a monumental work of art worthy of Christo. We humorously reflected on how we should submit a grant to the National Endowment for the Arts to raise additional funds for the project.

By March 14, Rodolfo had returned. He and his crew now faced a pressing problem: we had to get the skeleton out of the ground before the field season ended, and we had only two weeks to do it. This work would involve several phases. First, we would have to remove the rocks overlying the skeleton and open up a large quarry to have sufficient space to work. Then, trenches would have to be excavated around the skeleton. Because the skeleton was too large to be lifted out of the ground as a whole, we had to decide where to cut between the bones so that it could be divided into smaller, more manageable blocks. But before these blocks could be cut and lifted, the skeleton would have to be covered with toilet tissue and plaster bandages to form a protective covering and reinforced with sturdy wooden struts to hold it together and support it during the trip back to the museum. Once this was completed, we would have to define the different blocks by cutting through the rocks and, unfortunately, through some of the bones. Then all of the blocks would have to be undercut so plaster bandages could be attached to the undersides. These bandages would keep dirt and fossils from falling out of the plaster jackets when they were flipped over in the quarry so that the bottoms of the blocks could be jacketed. Finally, we would have to build wooden pallets for the blocks to sit on and find a crane to take them from the quarry to the road, where they could be lifted onto a truck for the trip back to the museum.

All of that work kept Rodolfo, Alberto, Gerald, Anwar, and several other crew members busy for the rest of the field season. We decided to subdivide the skeleton into five blocks: a large one for the hips and tail, two smaller ones for the hind limbs and arms, another large one for the trunk and neck, and a last one for where we hoped the skull would be. Even after dividing the skeleton into five sections, some of the blocks weighed more than a ton with their petrified bones, wet plaster, surrounding rock, and wooden supports.

Meanwhile, along with Luis, Lowell and the other members of the geological team embarked on a two-day trip to the west of Auca Mahuevo, in search of a volcanic ash described in a geologic paper we had read. The scenery on the drive was spectacular; huge volcanoes loomed on the horizon off to the west. But as we approached the site where the ash had been discovered, the dirt road was washed out, so we could not get to the precise locality described. After driving almost completely around the site on other roads, however, we did find one track that got us close. At the end of the track were some exposures that looked as if they contained the same sequence of rock layers exposed at Auca Mahuevo. Since the sun was setting, we decided to cook dinner and look at the exposures in the morning.

We bedded down soon after dinner, but we were awakened by vigorous thunder and lightning about 1:30 A.M. The storm was rapidly approaching, and within a few minutes the first sprinkles began to pelt our sleeping bags. We hadn't brought our tents because there wasn't room in the vehicle, so as the rain quickly increased, we crawled out of our bags, stuffed them in or under the vehicle, and scrambled inside. With five of us and some of our equipment, it was pretty cramped, and for a few minutes we couldn't get the power windows to close. By the time we finally managed to close them, most of us were soaked. Heavy rain and a bit of hail pummeled our vehicle, and it took about an hour and a half for the storm to pass. But by three or three-thirty, we slumped out of the car, laid out our air mattresses, and crawled back into our bags.

Arising early in the morning, we ate and then assaulted the exposures. Near the top of a large ravine, Lowell found a cream-colored layer of rock that was quite different from any of the others. It seemed to be a layer of altered volcanic ash, but there didn't appear to be any large crystals in it that might be used for dating through radioactive methods. Nonetheless we collected some plastic bags full of chunks, hoping that not all of the crystals had weathered to clay. Luis also found a few fragments of fossil bone, but nothing worth collecting. We would need to return some other time and try to get to the site described in the scientific paper, but that would probably require hiring either horses or a helicopter. At about noon, we headed back to camp. As we drove, we passed drifts of hailstones that had washed into nearby gullies. Although we had spent a miserable night exposed to the elements, we clearly had not borne the brunt of the storm.

We returned to find that Frankie, Gerald, and the rest of the crew had found three embryos in the eggs at the quarry. In addition, they had measured and mapped another area of eggs on the flats. In the first large grid, seventy-four clusters had been documented, whereas the second, smaller grid contained about half that number.

Over the next three days, most of the paleontological team focused on plastering and flipping the plaster jackets containing the abelisaur skeleton. Meanwhile, Lowell's geological team began looking closely at some of the exposures that contained eggs in the highest layer in the sequence. Brushing off the loose dirt, we could see that the eggs had been laid on a bumpy surface of the ancient floodplain. Assisted by Julia and Frankie, Lowell used a tape measure and the leveling bubble in his Brunton compass to map where the eggs sat on the surface. The surface of the ancient floodplain was definitely bumpy, with small mounds and depressions, just as David Loope had warned us to expect with vertisols. However, it did not appear that the eggs sat in these small depressions. We wondered why the depressions had not been used as nests, but we could not be sure. The best news was that, at last, the swarms of mosquitoes were now quickly dying off.

Most of our last five days in the field were devoted to moving the blocks containing the abelisaur skeleton out of the quarry and down to the nearest road, where they could be lifted onto a flatbed truck for the trip back to Rodolfo's museum. The work went slowly, as some of the blocks weighed over two thousand pounds. To flip these large rocks and move them around the quarry, we used our four-wheel-drive Passport that American Honda had given us to use during the expedition. After all the blocks were flipped and ready to move, our friends at a nearby oil station helped us find a crane, and Rodolfo arranged for a truck to transport the blocks to Plaza Huincul. Although the truck failed to come on the specified day, Rodolfo returned a couple of days later with the driver to pick up the plaster jackets that the crane had left by the side of the road. In addition to this work, we also entertained several individuals who had helped sponsor our expedition, as well as several members of the media both from local and international networks and publications.

Finally, on the afternoon of March 24, we packed up most of our gear and celebrated the success of our season with a final *asado*. The next morning, under once again rainy skies, we headed for Plaza Huincul on our way back to Buenos Aires. It had been another exceptional month of discoveries. But now, once again, we faced long months of preparing the fossils that we had collected and analyzing the data that we had gathered.

CHAPTER ELEVEN

Was the Nesting Site
Used More than Once?

Discovering New Egg Layers

Our primary goal in undertaking the second expedition was to gain a better understanding of the reproductive behavior of sauropods. Given the evidence that was preserved in the rocks at the site, the answer to one aspect of this behavior had become clear. Many modern animals return to the same area to lay their eggs during different breeding seasons. Did the ancient giant dinosaurs of Patagonia do this also?

The evidence to answer this question was in the stratigraphic section that showed all of the different rock layers exposed at the site. To make the drawing, Lowell began at the lowest rock and worked his way up, layer by layer, to the highest. He had taken great care, when compiling the section, to record the color, thickness, rock type, and fossil content of each layer of rock that he found, not the most exciting job in the world even for a geologist. In all, the layers of rock that crop out at Auca Mahuevo are almost five hundred feet thick, and there are more than thirty-five distinct, major rock layers. The project took more than a week, but in the end, this hard and sometimes tedious work paid off. With all this geologic data recorded in his notes, we could now determine how many different layers of rock contained sauropod eggs, which would provide the clues we needed to decide if the sauropods had used this nesting site more than once.

As we mentioned earlier, this sequence of rock layers is important for telling time back when sauropods roamed the ancient Patagonian landscape. In a sense, each layer represents a page in the book of geologic history for Auca Mahuevo. By applying the geologic principle of superposition, we can leaf through that book page by page, from older, lower layers into younger, higher layers. Eggs and embryos found in the lowest rock layers were laid at some time before eggs

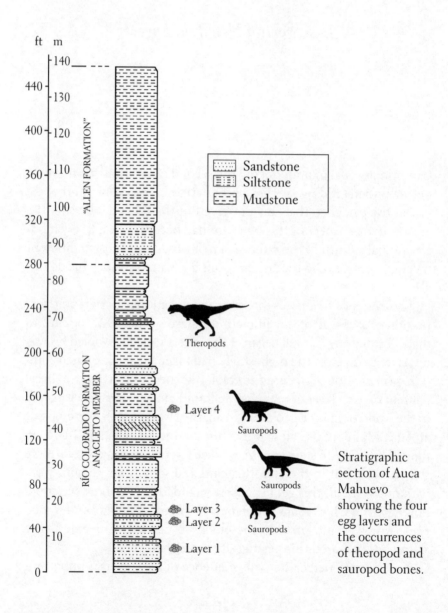

Stratigraphic section of Auca Mahuevo showing the four egg layers and the occurrences of theropod and sauropod bones.

found in higher rock layers, which is why it was so important for us to determine how many different rock layers contained eggs. Each egg-bearing layer represents a different nesting event because each rock layer represents a different page or moment in the history of the site. So, if we found sauropod eggs in more than one of the rock layers at Auca Mahuevo, we could be certain that the sauropods had laid their eggs more than one time at the site.

During our first trip to Auca Mahuevo in 1997, we had discovered that the eggs on the flats and the eggs in the embryo quarry both came from a twenty-four-foot-thick layer of mudstone that began about fifty feet above the lowest layer of rock exposed at the site. At first, we weren't sure whether the eggs from the flats and the eggs from the embryo quarry about a mile away actually came from the same layer. However, after hiking on the mudstone layer that contained the embryos all the way back to the flats, we were sure that all the eggs came from the same mudstone layer, leading us to conclude that there was one enormous nesting site covering a couple of square miles. That conclusion still stands, and our subsequent investigations showed the area to be even larger. The evidence we discovered during later trips to the site, however, documented that the sauropods had laid their eggs at Auca Mahuevo several different times.

We got our first indication of this in May 1998 before our second expedition. Lowell, Luis, Rodolfo, Suzi Zetkus, and Pablo Puerto had briefly returned to the site because *National Geographic* was planning to do an article on our discoveries. Brooks Walker, a photographer working for the magazine, accompanied us throughout our weeklong stay. Most of our time was spent helping Brooks take photos around the flats and at the embryo quarry for the article, rather than conducting scientific work. But one day, late in the week, Brooks climbed up to the top of a small ridge overlooking the site to get some panoramic shots. The ridge was across the road to Auca Mahuida from where we had found the eggs on the flats and the embryo quarry. Consequently, we had not spent much time looking for fossils there yet. As Brooks was taking pictures, he noticed some eggs weathering out of the ground near the base of a small peak on top of the ridge. After finishing his photos, he came back down and told us about his discovery.

Of course, we were thrilled to learn that more eggs were present in a new area of the site. We suspected at first that the eggs were coming out of the same layer of mudstone that had contained the eggs on the flats and at the embryo quarry, but we had to check out Brooks's new site to be sure. As we walked east about a half mile from the road to the ridge, it soon became clear that the eggs Brooks had found were at a much higher elevation than the eggs on the flats. We knew that more hiking would be required to determine where the new layer fit in the sequence of layers, especially in relation to the layer that contained the eggs on the flats, but we would have to wait until our next expedition to conduct that detailed exercise.

One of our primary geological goals when we returned to Auca Mahuevo in March 1999 was to make sure that Brooks's eggs came from a new layer. To do that, Lowell, Julia, and Alberto traced the layer of mudstone that contained the eggs on the flats to the east along the base of the ridge where Brooks had found the new eggs. The egg-bearing layer on the flats ran right along the bottom of the ridge, whereas Brooks's egg site was situated near the top. More than seventy-five feet of mudstone and sandstone layers separated the two layers containing eggs. Consequently, we knew that the sauropods had laid the eggs preserved on the flats long before they had laid the eggs at Brooks's site.

At present, we cannot tell how long the eggs on the flats were laid before the eggs at Brooks's site. It's not likely that that much sediment could have been deposited within a few years or decades, but whether that seventy-five feet of rock took centuries, a few millennia, a few tens of thousands of years, or even a hundred thousand years to accumulate is unknown. Because we have not found any ancient layers of volcanic ash at the site to date through radioactive means, we just cannot be sure. We could not use the magnetic information contained in the rocks to make such estimates because all the rocks we have sampled thus far at Auca Mahuevo were deposited during an interval when the earth's magnetic field was reversed. Nonetheless, because of the law of superposition, we can be sure that the eggs contained in the mudstone layer at the flats and the embryo quarry represent a distinctly earlier page in the geologic history of Auca Mahuevo than the page represented by the layer containing Brooks's eggs.

It might seem a bit strange for a photographer to discover an important new locality laden with fossil dinosaur eggs. That was obviously not the primary reason why Brooks had come with us to Patagonia. However, he had followed us around the site for several days, and during that time he had developed a clear idea of what we were looking for. So when he saw the eggs on top of the ridge, he knew exactly what they were.

Professional photographers often accompany scientific expeditions, and Brooks was not the first photographer accompanying a fossil-collecting expedition to make an important discovery of dinosaur eggs and nests. Another such incident had happened about seventy years before in the Gobi Desert of Mongolia. During the 1920s and early 1930s the photographer J. B. Shackelford had documented the expeditions to Central Asia led by the famous explorer Roy Chapman Andrews and the paleontologist Walter Granger of the American Museum of Natural History. Like Brooks, Shackelford became finely tuned for fossil prospecting, and like Brooks, Shackelford's acute vision and attention to detail paid great dividends to Andrews and Granger. Andrews described Shackelford's moment of discovery in his account of the expedition:

> My car was far in advance of the others and I asked Shackelford to stop the fleet while I ran over to the yurts for a conference with the inmates. During this time . . . he wandered off a few hundred yards to inspect some peculiar blocks of earth which had attracted his attention north of the trail. From them he walked a little farther and soon found that he was standing on the edge of a vast basin, looking down on a chaos of ravines and gullies cut deep into red sandstone. He made his way down the steep slope with the thought that he would spend ten minutes searching for fossils and, if none were found, return to the trail. Almost as though led by an invisible hand he walked straight to a small pinnacle of rock on the top of which rested a white fossil bone. Below it the soft sandstone had weathered away, leaving it balanced and ready to be picked off.
>
> Shackelford picked the "fruit" and returned to the cars, just as I arrived. Granger examined the specimen with keen interest. It was a skull, obviously reptilian, but unlike anything with which he was familiar. All of us were puzzled. Granger and Gregory named it *Protoceratops andrewsi* in 1923. Shackelford reported that he had seen other bones, and it was evident that we must investigate. . . .

Granger brought in . . . a part of an eggshell which we supposed was a fossil bird, but which subsequently was recognized as dinosaurian. . . . We could hardly suspect that we should later consider this the most important [site] in Asia, if not in the entire world.

The skull that Shackelford had found turned out to belong to a primitive new member of the horned dinosaurs, and the eggshell that Granger had found following Shackelford's route led to the discovery of numerous nests of dinosaur eggs. These were the ones that Andrews and his team originally thought belonged to *Protoceratops*, but were later recognized as belonging to a dinosaur related to *Oviraptor*.

Having determined, with Brooks's help, that more than one layer of eggs was present at the site, we now needed to make sure that no others were present. We examined the layer containing the eggs on the flats in more detail. Was there really just one thick layer or were several layers closely packed together? In March 1999, Lowell and Julia Clarke spent two whole days walking around the flats and the ridges that rimmed them. This careful investigation revealed two separate layers of eggs in the mudstone, separated by four thin layers of sandstone, siltstone, and mudstone that together were about five feet thick. The lower layer did not contain nearly as many eggs as the upper layer, but it represented a different and slightly earlier page in the history of the site. Our survey of the flats also showed that both of these layers of eggs were well below the younger layer that Brooks had found. Perhaps these two layers of eggs were laid within years or decades of one another, but again we cannot be certain.

In addition, Lowell found another single cluster of eggs in a fourth layer about twenty-five feet below the lowest layer of eggs exposed on the flats. Not much of this layer is exposed because it lies right on the desert floor below the ridges and flats. Nonetheless, we know that the eggs were not washed down from the adjacent egg-bearing areas because they are complete and planted firmly in the ground, rather than sitting isolated on the surface.

Finally, Frankie's mapping of eggs at the quarry, along with statistical analyses of their distribution, which we will describe later, suggested that two different levels of eggs were probably present there. This means that at least four, if not five, different layers of sauropod eggs are present at Auca Mahuevo.

In order to identify the main four egg layers separated one from another by several feet of rock, we decided to number them in ascending order. We called the layer containing the isolated clutch Lowell had discovered twenty-five feet below the lowest layer on the flats egg layer 1. The lowermost egg layer at the flats was labeled egg layer 2. The egg layer containing our quarry and all the eggs we had found during our 1997 expedition was identified as layer 3. Finally, the uppermost egg layer that photographer Brooks Walker had first spotted was called egg layer 4.

That four distinct rock layers contained clusters of sauropod eggs at Auca Mahuevo provided important evidence about the reproductive behavior of these huge dinosaurs. Based on the principle of superposition, these giant sauropods clearly returned to the nesting site at least four different times to lay their eggs, a behavior called site fidelity. Since we don't know the precise age of each rock layer, we don't know whether they returned every year, but that is certainly possible given the probability that not all of the eggs laid by the sauropods at the site were preserved as fossils. In years when floods did not bury the eggs quickly, most of the embryos would have hatched and left the site. The eggshells would have been broken up, and the shell fragments would probably be dissolved by rain or destroyed by other natural processes, leaving no evidence of these breeding events in the fossil record. So, while we cannot be certain exactly how many times or how often the sauropods used the site, we do know that they used it at least four or five separate times.

Because the surface ornamentation of the eggs varies, both within each layer and between different layers, we wondered whether the same kind of sauropod had laid all the eggs. But after careful study in the lab, we concluded that the differences in the microstructure are not so great as to suggest that more than one species of sauropod nested at Auca Mahuevo. Female dinosaurs, like all other egg-laying reptiles, laid slightly different kinds of eggs during their lifetime. For example, eggs laid by modern reptiles, including birds, tend to be larger in older individuals, and variations may occur depending on the environmental conditions and food available during a particular nesting season. Recent studies of modern mallard ducks has documented that females lay larger eggs when they mate with their preferred male partner. One theory is that the female invests more resources in the eggs resulting from a mate that she regards to be more fit than the others. We don't know whether sauropods also behaved like this, although it is possible since mallards and sauropods are both dinosaurs. Nonetheless, such factors may account for the variations we observed at Auca Mahuevo.

Additionally, we wondered how variations in environmental factors might come into play. As mentioned earlier, dinosaur eggshell is made of calcite, which is highly soluble in water. After the eggs were laid, dissolution by rainwater or subsequent exposure to groundwater after burial could have altered the surface ornamentation of the eggs, eroding or accentuating ridges and valleys. The eggs from Auca Mahuevo have been at the mercy of such conditions for millions of years, so it seems reasonable to expect that some dissolution could have occurred. So, until we discover an egg with a different kind of embryo inside or one with distinct differences in its microstructure, we think it is more reasonable to suggest that all the eggs from all the rock layers at Auca Mahuevo came from the same species of sauropod.

This was not the first time that paleontologists had suggested that sauropods returned to a particular nesting site. In 1995, for example, paleontologists working in the southern Pyrenees Mountains of Spain reported that they had found a site that also contained several clusters of dinosaur eggs in one layer of red sandstone. The eggs were almost spherical and slightly larger than the Auca Mahuevo eggs, averaging about eight inches in diameter. Although no embryos were found inside the eggs, they were assumed to belong to sauropod dinosaurs.

At this Spanish site, the organization of the sand grains within the layer that contained the eggs suggested that the sandstone had been deposited by waves along the shoreline of an ancient ocean. Although initial studies concluded that the dinosaurs had nested on this beach, subsequent examination of the rock layers containing the eggs demonstrated that the ocean had retreated from this ancient shoreline by the time the dinosaurs had laid the eggs. Fossils from the sandstone and other nearby rock layers indicated that these alleged sauropods had used the site sometime between about 71 million and 65 million years ago, making it slightly younger than our Patagonian site.

Across one area of about six thousand square yards, the collectors documented twenty-four egg clutches arranged in three clusters.

These clutches contained no more than seven eggs, significantly fewer than those at Auca Mahuevo. On average, the nests were about two to three yards apart from one another within the clusters, although the concentration of nests was not nearly as great as that discovered in Patagonia. Originally, researchers thought that the eggs were laid in depressions dug in the sand by the sauropods, but subsequent researchers have argued that one cannot be certain of this because there are no differences between the rock that entombs the eggs and the rock that underlies the eggs.

Other, smaller exposures of the sandstone layer containing eggs were found within a six-square-mile area surrounding the site. Based on the density of eggs at the best site, Spanish paleontologists estimated that up to three hundred thousand dinosaur eggs might be preserved in the red sandstone around the whole area. This led them to speculate that the sauropods nested gregariously, but this conclusion has been questioned by later studies of the eggs in this area. Other paleontologic teams working in this rugged region of Spain have identified several different layers of eggs that are similar in appearance and microstructure. These field studies indicate that, as at Auca Mahuevo, the alleged sauropods must have returned to the Spanish nesting ground multiple times to lay their eggs.

The other area of the world that has produced numerous dinosaur eggs in particular sites and many sites over a large region is in India. Large, round eggs, found in clusters that might represent nests, are preserved in rocks that formed near the very end of the Cretaceous period. These rocks are exposed in a region referred to as the Deccan Traps, a geologic province noted for its extensive lava flows that cover the southwestern corner of India. As in the case of the Spanish sites, the Indian eggs lack embryonic remains, but their large size has made paleontologists believe they were laid by sauropod dinosaurs. Some paleontologists have suggested that the large numbers of eggs discovered in this region are evidence that these putative sauropods returned to their nesting grounds multiple times. Although this is clear at both Auca Mahuevo and the Spanish site, the egg-bearing layers mixed in with the lava flows in the Deccan Traps are not as continuous. This has complicated attempts to match up the egg-bearing strata across the region and left some scientists skeptical about whether the Indian dinosaurs returned to their nesting grounds multiple times.

Based on the discoveries at Auca Mahuevo, however, we could now be certain that some sauropods returned to the same nesting site at least several times. But what was it like when they came to Auca Mahuevo to nest? Did the mothers come as individuals or in large herds? Did they dig nests or lay their eggs randomly across the floodplain? Once the embryos had hatched, did the mothers take care of their babies? We tried to solve some of these important questions by mapping the eggs in the quarry and on the flats.

Nancy Rufenacht prepares the mold of one of the nests found during the 2000 expedition.

Lowell Dingus and Julia Clarke map a vertical rock section revealing traces of the ancient topography of the nesting site.

The heavy plaster jackets containing the *Aucasaurus* skeleton are lifted by a crane for transportation to the Carmen Funes Museum in Plaza Huincul.

Building a large grid on the flats of egg-layer 3 for mapping the exposed egg-clutches.

Luis Chiappe, Lowell Dingus, and Rodolfo Coria (from left to right) at the 1999 main camp.

The bones of a titanosaur sauropod are carefully excavated during the 2000 expedition.

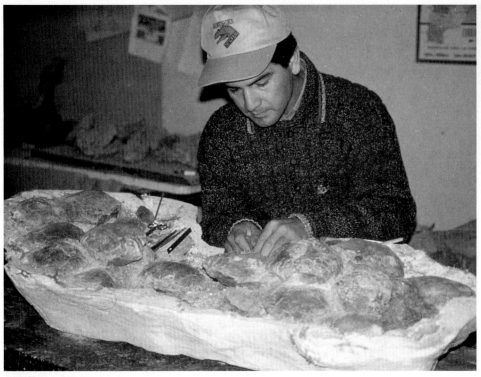

Carmen Funes Museum fossil preparator, Sergio Saldivia, removes the clay surrounding a clutch of eggs.

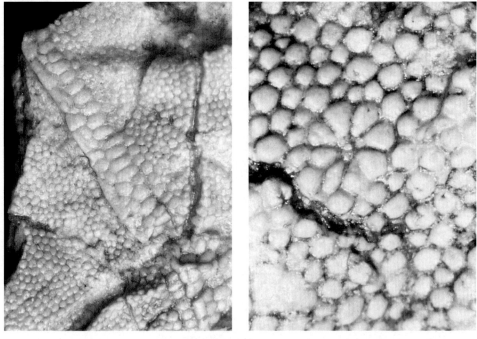

The embryonic skin shows a diverse pattern of scales. These unique fossils provided the first glimpse of the skin of dinosaur embryos.

Although somewhat crushed from rock compaction, the Auca Mahuevo eggs would have been round and 5 to 6 inches in diameter.

Skull and limb bones of an unhatched sauropod inside its egg.

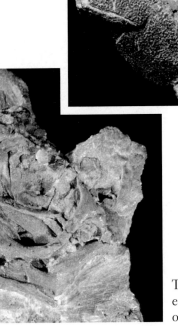

This nearly complete embryonic skull is only one and a half inches long.

The tail and hip of *Aucasaurus* is still encased by its plaster jacket.

Frankie Jackson maps
the distribution of the eggs
inside a nest.

Marilyn Fox examines eggshell fragments that fan out of a weathered clutch.

David Loope inspects a series of plate-like sauropod footprints.

Luis Chiappe excavates the remains of a titanosaur skeleton found in egg-layer 4 during the 2000 expedition.

The bared flats and hills where eggs were first discovered at Auca Mahuevo.

CHAPTER TWELVE

Were Giant Dinosaurs Gregarious?

Our Nests of Eggs and Other Evidence

Some animals are gregarious, whereas others live more solitary lives. How did the giant sauropods behave, especially when they went to the nesting site at Auca Mahuevo to lay their eggs? It had previously been suggested that sauropods laid their eggs in enormous colonies. The studies of the Indian and Pyrenees egg sites, mentioned in the last chapter, documented that thousands of eggs had been laid in areas encompassing several square miles. To the paleontologists studying those sites, this suggested that the area represented a single enormous nesting ground. In addition to speculating that the sauropods of India and the Pyrenees may have returned to the site during several breeding seasons, the scientists concluded that the close spacing of the nests might be indicative of some territorial behavior among individual sauropods. This seemed reasonable to the researchers because the eggs and nests were fairly well preserved, which suggested that the mothers had not trampled the clutches of eggs while moving around the site. But this evidence for colonial nesting and territoriality did not satisfy many researchers in the field of dinosaur reproductive biology.

Notwithstanding such criticisms, evidence from fossilized footprints and bone beds (high concentrations of skeletons from a single dinosaur species in a single rock layer) has also been used to suggest that at least some sauropod dinosaurs were gregarious. The best evidence for this comes from what are known as trackways, sequences of fossilized footprints left by sauropods in soft mud and sand, which later hardened into rock after the layer of soft sediment was buried.

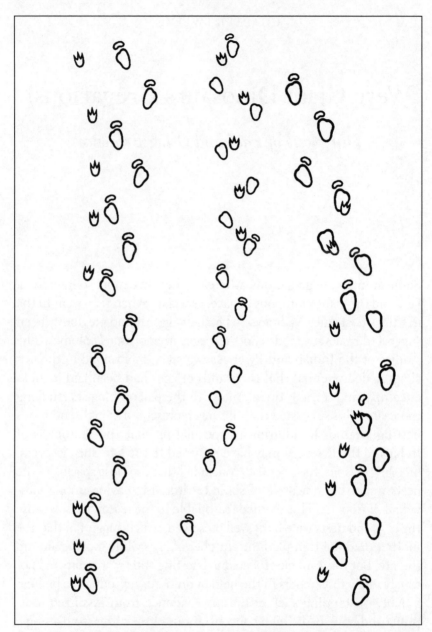

These dinosaur footprints preserved in over 100 feet of the Paluxy River bed in Texas show three sauropod trackways, each overlapped by theropod trackways (birdlike footprints). Discoveries like this one have led to the belief that sauropods traveled in herds and that large theropods may have stalked their prey in packs.

One of the early students of these dinosaur tracks was Roland T. Bird, a preparator and collector at the American Museum of Natural History. Additionally, Bird was one of the original "bikers." Throughout the 1930s and 1940s, Bird thundered across the American West on his Harley-Davidson motorcycle in search of dinosaur footprints.

Bird discovered numerous sites of dinosaur footprints throughout the Southwest, and in 1937 while riding across New Mexico and Arizona, he received a tip that gigantic fossil footprints had been found along the Paluxy River in central Texas. Upon arriving in the town of Glen Rose, he learned that the local inhabitants were well aware of the fossils. Eventually, Bird discovered the trackway of a large theropod dinosaur leading into the Paluxy River. With the help of a local work crew, he diverted the water and exposed the riverbed. Imbedded in the rock layer that formed the riverbed was an incredible sight. Fossil footprints documented that at least twelve sauropods, probably brachiosaurs, had walked in the same direction across a mudflat that bordered the ancient Gulf of Mexico. More incredibly, they had been followed at some later time by three large theropods. Analyses by Bird and later workers, including the foremost modern expert on dinosaur trackways, Martin Lockley from the University of Colorado near Denver, have shown that the theropod tracks overlap and impinge upon the sauropod tracks, documenting that the theropods passed after the sauropods. The sauropod tracks also show that the animals were walking side by side in the same direction at regularly spaced but substantial intervals apart from one another. Bird interpreted these trackways to mean that the theropods were stalking the sauropods and that perhaps one of the meat-eaters had even attacked one of the sauropods. In Lockley's analysis, however, no evidence in the theropod tracks suggests that the meat-eaters sped up to catch the sauropods. Nor does any evidence in the trackways suggest that the animals turned to fight one another. In fact, it is not possible to determine the time in between the passage of the two groups. It could have been moments, or even hours. The theropods may have been hunting the sauropods, but there is no evidence of an attack. Nonetheless, the evidence that the Paluxy sauropods were traveling in a herd seems convincing.

Bird discovered more compelling evidence of sauropod herding at the nearby Davenport Ranch in 1941. There, he discovered an outcrop

Several skeletons of oviraptorids, parrot-headed theropods from the late Cretaceous, have been found lying on top of their egg clutches in the Gobi Desert. These discoveries document that the brooding behavior of birds was inherited from their dinosaurian ancestors.

of rock containing the trackways of twenty-three sauropods all moving in the same direction within a narrow corridor about fifteen yards wide. The overlapping of the trackways, based on the analysis of Lockley, documents that larger sauropods were leading the way and younger, smaller individuals followed in line. The herd was moving at a modest walking pace, veering from right to left. There is no evidence, though, to suggest that the larger adults had encircled the smaller juveniles to protect them, as some researchers had speculated earlier.

Even though sauropods had a complex social structure that, at least at times, involved herding, it is still not known whether they gave their eggs and hatchlings parental care, that is, whether the adults fed and protected their young. Parental care has, however, been documented for other types of dinosaurs, including some meat-eating theropods and plant-eating ornithischians.

Working with another team of paleontologists from the American Museum of Natural History in the Gobi Desert of Mongolia, we were part of the crew that discovered the first fossil dinosaur skeleton—the parrot-beaked *Oviraptor*—actually sitting on its nest. Interestingly, when the first skeleton of *Oviraptor* was found in 1923, it was collected from on top of a clutch of eggs. When described in 1924, this was regarded as evidence for *Oviraptor*'s predatory activities. Paleontologist Henry Fairfield Osborn assumed that the *Oviraptor* had died while seizing the eggs of a plant-eating *Protoceratops*—a primitive horned dinosaur common in the Gobi deposits. Osborn's assumption led to *Oviraptor*'s condemning name, which means "egg seizer." But the eggs underneath that first *Oviraptor* skeleton did not contain any embryos, and their identity remained a mystery until seventy years later, when our crew discovered eggs of an identical shape and appearance that contained an embryo of an oviraptorid inside. Thus, this crucial piece of evidence showed that the skeleton collected in 1923 and those discovered decades later were actually brooding nests of eggs that contained their own kin. Subsequent discoveries by the American Museum team and by other paleontological expeditions of other Gobi skeletons in exactly the same pose provided solid evidence that certain types of dinosaurs, including *Oviraptor*, did care for their young, and evidence from other nesting grounds in Montana has suggested the same. The meat-eating dinosaur *Troodon*, one of the closest relatives of birds, has also been found to nest on top of its brood.

For the sauropods, however, this is not so clear. Some researchers believe that, because the eggs are laid so close to one another, the huge size of the adults would prevent them from directly taking care of their own young. Furthermore, the rarity of hatchlings or juveniles in the nesting grounds containing sauropod eggs is considered to suggest extreme precociality, meaning that as soon as the sauropod embryos hatched, they left the nesting area and moved to feeding grounds. Although we have found many embryos at Auca Mahuevo, we have not found any well-preserved hatchlings or juveniles in the same rock layer with the eggs. Anwar Janoo found a few small bones that might be from a sauropod hatchling on top of a clutch near the egg quarry in 1999, but they were too poorly preserved for us to identify them with any certainty. If the paucity of hatchlings and juveniles is a real representation of what was happening at the nesting site, and not a misleading problem with the way fossils became preserved, our inability to find juveniles may support the conclusion that sauropods were very precocial. One possibility was that, shortly after hatching, the baby sauropods congregated in large numbers, forming herds equivalent to flocks of hatchlings of flamingos and some other kinds of birds. These juvenile flocks, usually called crèches, are guarded by a group of adults. The same could have been true for the giant sauropods. This idea may explain why sauropod bone beds and herd trackways rarely contain fossils of hatchlings or small juveniles. In any case, we had plans to test this idea. And where better than in the richest sauropod nesting site yet discovered—Auca Mahuevo.

Our examination of the scientific literature revealed that most, if not all, of these ideas about the reproductive biology of sauropods had been reached with little supporting evidence. Speculation about colonial nesting and territoriality in sauropods had been based on a limited number of nests, and as we have discussed earlier, without even knowing the true identity of the eggs. We felt that Auca Mahuevo, with its extensive egg layers laid by one kind of sauropod, provided a unique opportunity for sampling and measuring the eggs at the site, which could in turn shed new light on these and other issues concerning sauropod nesting behavior. The crucial evidence to resolve these issues lay in how the eggs were distributed in the egg layers.

As mentioned earlier, we undertook two projects to determine the distribution of eggs at Auca Mahuevo. The first involved excavating a large quarry near the embryo site and mapping the position of each egg that was found there. We anticipated that this would give us a good idea about whether individual eggs were arranged in clusters that might represent nests or whether the sauropods had laid the eggs more randomly across the surface of the floodplain without constructing or utilizing distinct nests. Some evidence from other sites suggests that sauropods did the latter. At a site in southern France, for example, large eggs long thought to belong to sauropods appear to be laid in large semicircles without any indication that they were clustered in nests. At many other sites, megaloolithid eggs—the kind of eggs typically attributed to sauropods—have been found in more scattered patterns rather than in discrete clusters.

Frankie Jackson was instrumental in our egg-mapping at Auca Mahuevo. With the experience she had mapping dinosaur eggs in Montana, we knew that she should be the one to record the position of each egg in three dimensions. Our team surveyed the area within the quarry and laid out a system of one-meter squares, much as is done in an archaeological excavation. The initial quarry excavation covered twenty-five square meters, slightly more than twenty-five square yards. As new eggs were exposed during our excavation, Frankie would record their horizontal distances along the sides of the gridded rectangle from one corner of the quarry. We also recorded the elevation of each egg in the quarry with respect to a fixed point. The positions of more than two hundred whole eggs were eventually mapped, far more than in any other previous study.

After gathering the data, Frankie analyzed the maps back in the United States using a variety of statistical methods. To help her in these analyses, she recruited Richard Aspinall, a computer scientist at Montana State University in Bozeman. Frankie and Richard produced a three-dimensional map of our quarry that could be rotated and viewed from the top, the sides, and any other angle. The analyses showed the eggs were definitely clustered, not randomly distributed, although some eggs lay scattered between the more well-defined clusters, suggesting that the clutches may have been laid in nests of irregular shape. At least nine clusters were documented, and seven of those contained between fifteen and thirty-four eggs. This suggested that individual sauropods had laid clutches of eggs during one sitting. Given the way streams flow, we could not envision how currents during floods could have arranged the eggs in clusters like those we found in the quarry.

The number of eggs in a clutch also varies among modern egg-laying animals. The number usually correlates with the amount of food available during the nesting season, along with the abundance of parasites and predators. Ostriches usually lay between fifteen and thirty eggs; crocodiles are known to lay up to sixty eggs; and Komodo dragons tend to lay between twenty and thirty eggs. Thus, the large number of eggs in our clutches was not surprising. What was surprising, however, was that the number of eggs in a typical clutch was much larger than the number documented in suspected sauropod nests at other sites around the world, where there are usually less than ten eggs in a megaloolithid nest. Perhaps this significant difference reflects an abundance of food at Auca Mahuevo, a possibility that is also suggested by the immensity of the nesting colony.

The 3-D map also documented two separate egg layers in the quarry, separated by a few inches of mudstone, which suggested that

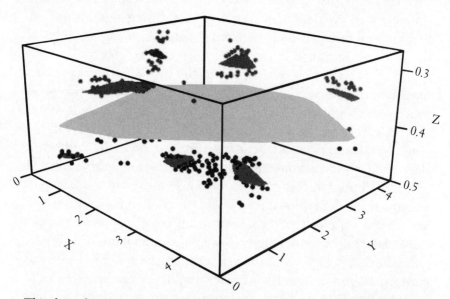

This three-dimensional map of eggs from layer 3 at Auca Mahuevo shows how these eggs are generally clustered in distinct clutches. The vertical scale of this diagram has been exaggerated to show better the existence of two levels of eggs, whose separation is depicted here by the intervening light gray plane.

two separate episodes of nesting had been preserved. Few if any eggshell fragments were found between the clusters, as we would have expected if the eggs had sat out on the surface for more than one year and the eggs had been broken through natural deterioration. Clearly, the eggs had been buried by flood debris before such breakage could occur. This lack of eggshell fragments also suggested that each layer of eggs represented a single egg-laying episode rather than an accumulation of nests that had been laid over several breeding seasons. At least several mothers had laid their eggs at this spot during a relatively short period.

The mapping at the quarry helped to answer another question involving parental care. We noticed that the eggs were generally complete and, therefore, unhatched, and as already noted, we observed relatively few fragments of eggshell between the egg clutches. All this suggested that the females did not remain at the site after they laid the eggs, because if they had, we would expect to find many more broken eggs and eggshell fragments as a result of trampling. As previously mentioned, strong evidence suggests that dinosaurs more closely related to birds, such as *Oviraptor* and *Troodon*, brooded their eggs in much the same way that birds do today. Some ornithischian dinosaurs may also have exhibited sophisticated parental care. Jack Horner and his colleagues from the Museum of the Rockies in Bozeman have suggested that hatchlings of the duckbill dinosaur *Maiasaura* were fed in their nest by their parents. Although evidence for this is not as conclusive as it is for *Oviraptor* and *Troodon*, it is believed that all dinosaurs offered some parental care to their broods.

Extinct dinosaurs belong to a genealogical group in which crocodiles and birds are the only living members, which means that the common ancestor of crocodiles and birds was also the common ancestor of extinct dinosaurs. Despite their fearsome appearance and predatory habits, crocodiles are exceptionally tender parents. After carefully burying her eggs in a sandy mound of vegetation, the mother guards the nest from intruders for about three months while the eggs incubate before hatching. When she hears the chirps of the babies emerging from the eggs, she delicately uncovers them, gently scoops them up in her mouth, and carries them to the closest pond or body of water. Birds have evolved a number of more intricate behaviors in parental care. Almost all of them incubate and protect the eggs

When hatched, baby sauropods would have been barely a foot long, smaller than the footprints of their parents. Several adults may have guarded the periphery of the nesting colony.

by sitting on the nest, and then they feed and defend the hatchlings. Because both crocodiles and birds care for their young, it is logical to infer that they inherited this behavior from their common ancestor. Furthermore, because this ancestor was also the ancestor of the extinct dinosaurs, most scientists agree that most dinosaurs must have provided their young with some kind of parental care. Based on our observations in the egg quarry at Auca Mahuevo, we ruled out the kind of parental protection in which parents directly care for their own nest, but less elaborate kinds of parental care may have taken place. Although it is impossible to prove, the sauropods from Auca Mahuevo could have communally guarded the whole nesting colony. Adults may have patrolled the periphery of the nesting area to ward off potential predators. We will never know for sure, but it seems unlikely that the eggs were left to the mercy of the fearsome predators that must have roamed the region at that time.

The quarry was also useful in documenting the spatial relationships between individual eggs, although it was not large enough to provide a lot of data on how the clusters were distributed across the nesting site. For that, we would need to document the position of clusters across a larger area. The flats where we had first discovered eggs, the better part of a mile away from the quarry, seemed to be the best place to do that because the eggs and clutches were weathering out on a large, relatively flat surface.

So Luis, Frankie, and Gerald Grellet-Tinner had returned to the flats and surveyed a larger grid in an area that seemed to contain a representative number of clusters. The rectangle that they laid out was about sixty-five yards long and thirty yards wide, with the whole area encompassing about two thousand square yards. The surface of the area was fairly flat. The difference between the highest and lowest point within the area was about two feet, which was not much more than the thickness of one clutch, so we assumed that only one layer of eggs was present. Within the mapped area, we found seventy-four randomly distributed egg clusters that we assumed represented distinct clutches. In a couple of spots near the middle of the area, clusters were packed together rather densely, only two to four feet apart from one another, but throughout most of the area, clusters were separated from each other by at least nine or ten feet. To double-check our observations, we constructed and mapped a second grid about three hundred yards from the first one. This area was less than half the size of the other, but our mapping showed the same high concentration of clusters: thirty-five clutches were found.

Given this evidence, what can we reasonably conclude about the nesting behavior of the Auca Mahuevo sauropods? First, the infrequency of broken eggs and eggshell fragments between the clusters of eggs argues that these are clutches laid by a single individual during one nesting. If we accept that, then we can conclude that fifteen to thirty-four eggs were laid by one mother during a single sitting.

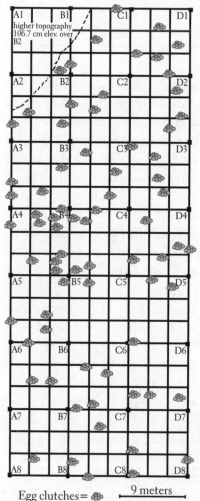

Egg clutches = 🐾

9 meters

This map of egg clutches exposed on the flat areas of Auca Mahuevo shows the high concentration of eggs present at this site. Maps like this one offer unprecedented evidence of the colonial nesting behavior of sauropod dinosaurs.

Second, although it is difficult to identify single nesting episodes involving more than one individual with complete certainty, the densely distributed egg clutches both in the quarry and on the flats suggest beyond any reasonable doubt that these sauropods were gregarious nesters. The alternative, that the hundreds of egg clutches exposed on the flats at Auca Mahuevo were laid at widely disparate times by solitary mothers, seems quite unlikely. Several birds and crocodiles congregate into groups—often, enormous ones in the case of birds—during the nesting season. The causes for this are not well understood, but scientists believe that nesting colonies offer better protection for the eggs, thus increasing their chances for survival. Perhaps this same adaptation drove the Auca Mahuevo sauropods to nest in large numbers.

Third, as we noted earlier, the sauropods returned to Auca Mahuevo numerous times, undoubtedly over many different breeding seasons, because at least four different layers of rock contain the same kind of fossil eggs.

What emerges from our investigations between 1997 and 1999 is a somewhat fuzzy, almost impressionistic portrait of the sauropods' life on the ancient floodplain during their breeding season. It is likely that herds of gravid females lumbered to Auca Mahuevo during many, if not all, of the breeding seasons in which the nesting site was in use. It is not clear whether males accompanied them. Upon arrival, each female laid between fifteen and forty eggs in a nest on the gently sloping, hummocky ground away from the major stream channels. Most of the nests were spaced five to fifteen feet from one another. Sometime after laying the eggs, the females probably left the site, leaving the eggs to incubate, although some adults may have remained in the area to guard the colony. The embryos grew inside the eggs to a length of about twelve inches before they were ready to hatch. At the end of their development inside the eggs, the embryos began to exercise their jaw muscles and to grind their teeth, in preparation for eating vegetation when they hatched. Normally, many of the eggs hatched without incident, and the hatchlings, with their relatively large heads, strong jaw muscles, and small but fully developed teeth, immediately set about consuming any nourishing vegetation that they could find. Over the next fifteen to twenty years, assuming they survived, the sauropods would increase in body size more than thirtyfold

Our discoveries documented that sauropods gathered in large numbers to lay their eggs in the soft substrate of a floodplain. Occasional heavy rains brought floods to the nesting colony, burying and preserving the eggs and embryos of these large dinosaurs.

to become some of the largest animals known to have walked the earth. However, floods occasionally inundated the nesting site during incubation, burying the eggs under a layer of mud and killing the developing embryos inside.

Except for the occasional carnage caused by the floods, this portrait of the nesting scene seems rather peaceful and pastoral. But based upon another discovery made during the 1999 expedition, that perception is misleading. A menacing terror lurked in the shadows of the floodplain.

CHAPTER THIRTEEN

An Awesome Predator

Discovering a Skeleton
of the Sauropods' Adversary

One might think that being fifty feet long and weighing several tons would make adult sauropods, such as those that lived around Auca Mahuevo, quite invincible. But a pilgrimage to the nesting site at Auca Mahuevo was clearly not without risks, even for giant sauropods and their offspring. Fearsome predators roamed the floodplain.

During the first expedition to Auca Mahuevo, no skeletons of adult dinosaurs were found at the nesting site despite several efforts to find some. That we had found only a few fragments seemed quite curious to us: With all of the eggs and embryos, why were there no fossils of adults?

That all changed with Alberto Garrido's discovery of the large carnivorous-dinosaur skeleton that we described while recounting events during the 1999 expedition. The portions of the skeleton that we had uncovered showed that the new dinosaur was either a smaller version of *Carnotaurus* or one of its close abelisaur relatives. We estimated that it was about 70 percent as large as the only known skeleton of *Carnotaurus*, yet based on the structure of the bones, we knew definitely that it was an adult. A number of questions kept flying through our minds as the long days of excavation passed. Could it simply represent a different sex than the other specimen of *Carnotaurus*? Could it be a specimen of *Abelisaurus*, another abelisaur known only from a skull that was collected from the same layers of rock about one hundred miles away? Could it be a previously unknown species that was

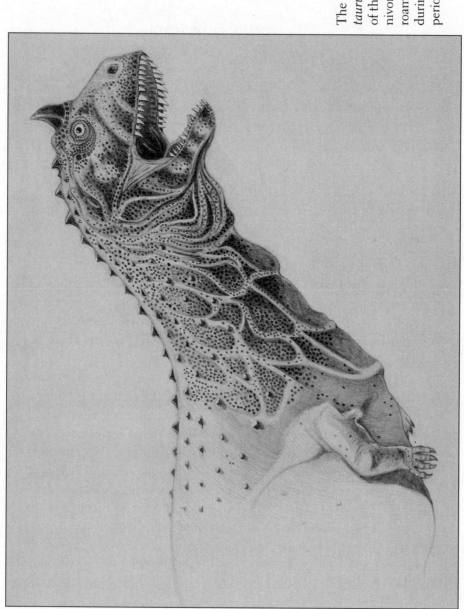

The 25-foot-long *Carnotaurus* is a horned member of the abelisaur family, carnivorous dinosaurs that roamed South America during the late Cretaceous period.

smaller but similar to *Carnotaurus?* To decide, we would have to transport the skeleton back to the lab and clean up the bones because the last and most crucial piece of evidence needed to establish its identity, the skull, was still missing. The few fragments of skull bones that we had uncovered were quite troubling because they suggested that the skull might have been destroyed. But perhaps the skull had just broken away from the rest of the body and was buried nearby in the hillside.

Because we had been unable to locate the skull, we had decided to take a large block of mudstone from the area where the skull could have been buried, hoping to find it inside when the block was prepared at the museum. Although our hopes were not too high, a couple of months after we returned home, Rodolfo called with great news. His chief preparator, Sergio Saldivia, had found most of the skull in the block. At least one side of it was fairly complete, which would allow us to make more concrete comparisons with other abelisaur skulls and determine its identity.

When the skeleton was fairly clean, Luis and Rodolfo started comparing it to other abelisaurs. Abelisaurs are primitive carnivorous dinosaurs known primarily from the late Cretaceous of the Southern Hemisphere. Most species, as well as the best-preserved specimens, of this theropod family come from Patagonia, including *Carnotaurus*, *Abelisaurus*, and *Ilokelesia*, but abelisaurs are also known from the late Cretaceous of India and Madagascar. Fragmentary remains have been reported from Western Europe, but these are poorly preserved and inconclusively identified.

Although the shapes of its bones showed that our predator from Auca Mahuevo was closely related to *Carnotaurus*, it was distinctly different from that species. Compared to the skull of *Carnotaurus*, our skull was proportionally longer but not as tall. In addition, *Carnotaurus* has large, prominent horns on the skull above the eyes, whereas the Auca Mahuevo abelisaur had only small bumps. As mentioned earlier, the arms of the Auca Mahuevo abelisaur were proportionately longer than those of *Carnotaurus*, although they were still quite short and reduced in relation to those of most other meat-eating dinosaurs. The bones of this skeleton were superbly preserved, but our biggest surprise was finding small, fossilized casts and impressions of the predator's muscles preserved above the hips in the mudstone that surrounded the skeleton. The most important aspect of the new skeleton was its completeness. Only the end of the tail, along with part of the skull, was missing; the hands and feet were essentially complete, giving us our first look at what the entire arms and legs of abelisaurs looked like.

All of this put us in a fortunate position. After analyzing the bony details preserved in the skeleton, we knew that Alberto had discovered a completely new species of dinosaur, which required a new scientific name. Coining a new name is part of describing a new dinosaur for the scientific community. After some consideration, we named the new abelisaur *Aucasaurus garridoi* to commemorate two of its attributes. The name for the new genus, *Aucasaurus*, means that it is a new dinosaur from Auca Mahuevo, while the name for the new species, *garridoi*, celebrates that it was discovered by Alberto Garrido. Alberto had not only found the specimen, but had also done much of the hard work required to collect it, so it seemed most fitting to recognize his extraordinary efforts.

About one year after the specimen was discovered, all of the jackets had been prepared. Rodolfo and Luis had made all the essential

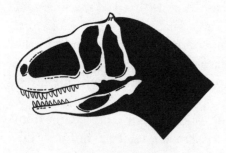

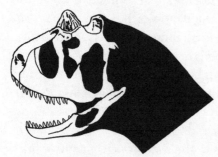

Although very similar in many aspects, the abelisaur *Aucasaurus* had a skull that was longer and lower than that of its close relative, the horned *Carnotaurus*. These two abelisaur theropods are the most completely known meat-eating dinosaurs discovered in Cretaceous rocks of the Southern Hemisphere.

The abelisaur *Aucasaurus*, whose skeleton was unearthed in our 1999 expedition, follows the track of a sauropod. Although closely related to *Carnotaurus*, the smaller *Aucasaurus* lacked the more prominent horns of its relative.

comparisons with other known dinosaur skeletons, and we were ready to write a scientific paper to announce our discovery. The article not only named the new abelisaur but also described several important anatomical features that had previously been unknown in abelisaurs. *Aucasaurus* is the most complete abelisaur skeleton ever collected and provides new insights into the evolution of this peculiar lineage of horned carnivores. Our paper was submitted to the primary scientific journal that publishes information about new research in vertebrate paleontology.

Beyond describing and naming the new dinosaur, we were still faced with one other mystery concerning this individual. How had it died? Once again, some clues were preserved in the rocks that entombed the specimen.

The rock that had produced the skeleton was unusual because no other layers at the site were quite like it. The skeleton was found in finely banded, purplish gray mudstone, and upon closer inspection, Lowell discovered that these laminations also contained small fossils of shelly invertebrates. Finely banded layers of mudstone are often deposited on the bottom of shallow lakes, ones that could easily have formed on the floodplain after storms. Similar layers can be observed forming on the bottom of many lakes today. We could once again use a geologic process that we see operating today to interpret how the banded mudstone formed 80 million years ago at Auca Mahuevo. Clearly, the carcass had been buried at the bottom of a shallow lake on the floodplain. Either the predator had died in the lake or its carcass had floated out into the lake before sinking to the bottom and being buried. Because part of its skull was broken apart, perhaps it had gotten in a fight and been killed when another dinosaur bit or struck its head. But at this point, we simply cannot be certain about the cause of death. Only further study of the bones might help answer that mystery.

CHAPTER FOURTEEN

The Y2K Expedition

With two successful expeditions to Auca Mahuevo, we had learned a lot about the reproductive behavior of sauropod dinosaurs, but many questions still remained open: What kinds of sauropods were responsible for these amazing nesting seasons? Could we find evidence to fine-tune our identification of them? Was the egg-clutch distribution of the extensive egg layer 4 comparable to the highly concentrated and randomly distributed clutches from egg layer 3? Were the sauropods laying their eggs in natural depressions, on flat surfaces, or in holes they purposely dug? These important questions and the many possibilities for finding exciting new dinosaurs in the fossil-rich rocks of Auca Mahuevo prompted another large-scale expedition in March 2000.

Luis and Lowell procured additional funding from the National Geographic Society and the InfoQuest Foundation, as well as the Ann and Gordon Getty Foundation in San Francisco and the Phillip McKenna Foundation in Pennsylvania. The Fundación Antorchas in Argentina awarded the project a generous grant, which allowed us to both expand the expedition and conduct cooperative post-field research between Argentine and American paleontologists.

Our Y2K expedition started like the others. A large component of our crew flew to Buenos Aires at the beginning of March, and soon after, they embarked on the daylong ride to Auca Mahuevo. Others flew directly to Neuquén to buy supplies, later meeting those leaving from Buenos Aires at the main camp. Rodolfo Coria and his team of fossil hunters would meet us at Auca Mahuevo. Minor issues such as losing bags and missing flight connections aside, this regrouping worked well, and by March 5 we were setting up our camp under the dim, reddish light of dusk.

The twenty or so people who composed our crew included the usual suspects. The team of paleontologists included Rodolfo and his staff from the Carmen Funes Museum, our egg specialists Frankie Jackson and Gerald Grellet-Tinner, and a number of technicians, students, and volunteers from Argentina and various other countries. Several new fossil hunters and geological specialists also joined us. The international flavor of our expedition team continued to expand, with the addition of a young Italian student, Giuliana Negro, whose warm temperament and enthusiasm for dinosaurs gained her a handful of joyful nicknames. Nick Frankfurt, our illustrator for this book, came along to get the kind of firsthand experience that all artists want when working on a project. He also assisted with some paleontologic activities. The entire team benefited from the superb cuisine of our cook, Omar Garcés, whose diligence and cordiality made our camp a much more pleasant place.

After we finished setting up camp on March 6, we resumed many of the tasks we had intended to continue when we had left the year before. Luis, Frankie, Gerald, and others returned to our quarry in egg layer 3 and began removing the overlying rock on some one hundred square feet of surface. We wanted to augment our map of eggs and to explore the relationship between them and the slickensides we had found the previous year. This season we were much better prepared for large excavations because we had brought a portable air compressor and pneumatic tools that enormously speeded the removal of the three feet of sterile rock covering the egg layer. These tools also saved us from painful evenings of recuperation after the backbreaking work.

Before completely exposing the egg layer, we decided to map on our egg-distribution chart the numerous slickensides that crisscrossed the quarry surface and intersected each other at varying angles. We marked the slickensides with fluorescent spray paint to help us understand both the origin of the slickensides and the distribution of the eggs. Luis, Frankie, Matt Joeckle, and David Loope spearheaded this operation. A reserved man with a cynical sense of humor and an expert on fossil vertisols, Matt examined the mudstone containing the eggs to help us understand the climatic conditions that had predominated at the time the dinosaurs nested. Once highlighted with green fluorescent paint, the distribution of the slickensides convinced Matt that the mudstones entombing the eggs did indeed represent vertisols, which was important in understanding the ancient climate of this site. In modern settings, vertisols form under semiarid conditions, and we can infer that the same environmental conditions prevailed in places where vertisols formed during ancient times.

The delicate task of uncovering the eggs took days of tedious work. Luis, Frankie, Gerald, and other team members spent long days under the merciless sun while gusts of wind blew grit into their eyes. Their task was complicated by the slickensides, which had taken a significant toll on the preservation of the eggs, often flattening them like pancakes. Even so, we were all excited to observe this ancient process of soil formation so intimately. Seeing eggs that had been displaced a foot or so deeper from the original level of their clutch gave us a much clearer idea of the extreme plasticity of this sediment. Naturally, Matt was more excited than anyone else because soil experts rarely have the opportunity to study extensive exposures of ancient soil in three dimensions, much less when they are packed with dozens of 80-million-year-old dinosaur eggs.

During resting breaks at the quarry, some of us prospected in the adjacent hills for fossilized patches of embryonic skin. Particularly successful at this was Adrian Garrido, a young and quiet technician at the Carmen Funes Museum (unrelated to Alberto Garrido), who found several beautiful patches of fossilized skin in clutches near our egg quarry. His finds were particularly important because all the skin we had previously found was from the flats, a mile or so from our egg quarry. Adrian's discovery indicated that the heavily cemented egg fragments of the flats that contained the skin of unhatched sauropods were also present around the quarry. Unfortunately, we did not have the pleasure of having Natalia Kraiselburd, our ace collector of fossilized skin, on our 2000 expedition because she had taken a new job early in the year. Adrian and others, however, picked up the slack by finding some remarkable skin patches. Particularly productive was one morning in which we all went to look for skin at the flats.

Lured by the beautiful specimen of the new abelisaur that we had found the previous year, a good portion of our team, headed by Rodolfo, spent days walking over the naked badlands of Auca Mahuevo in search of more fossil skeletons, not an easy task under the scorching sun. The first moments of excitement came on March 10 when Rodolfo and some of our fossil hunters found a few bones of a meat-eating dinosaur, which, although fragmentary, documented the presence of a colossal predator, much larger than the twenty-foot-long *Aucasaurus*. One of the fossils represented the end of a hipbone called the pubic boot. The similarity in shape and size of this fragment to those of the older *Giganotosaurus*, one of the largest carnivorous dinosaurs ever discovered, suggested that a relative of this fearsome creature had survived to roam Auca Mahuevo's river plains. Since Rodolfo had originally studied and described the colossal *Giganotosaurus*, the glimpse of another such beast waiting to be discovered at Auca Mahuevo fueled him and his team with a burst of determined excitement throughout the next several weeks.

That day of prospecting also led to a second important discovery: another flat area exposing large numbers of egg clutches. This surface was somewhat smaller than that of the flats at egg layer 3, but it was a stratigraphically higher part of our uppermost layer of eggs, egg layer 4. We knew that this would allow us to map the distribution of egg clutches in egg layer 4 without having to remove huge amounts of overburden, as well as to compare the distribution of eggs in layers 3 and 4. It was yet another day for celebration. Back at camp, our cook, Omar, welcomed us with cold drinks and snacks that were followed by a sumptuous *asado*.

Equally exciting discoveries were made in the following days. On March 12 two partial skeletons of titanosaur sauropods were spotted. As we mentioned earlier, we had found skeletons of these beasts in 1997 near Doña Dora's *puesto*, but this time, the bones were in exactly the same rock layers as the eggs of egg layer 4. Andrea Arcucci, an Argentine paleontologist from the Universidad de San Luis in central Argentina, found one. Andrea is a specialist on South American Triassic reptiles and a seasoned field person. Chatty and humorous, Andrea was highly missed when she left a week before the end of the expedition. Years of strolling across the reddish badlands of Triassic rocks in Argentina had given Andrea a keen eye for discovering fossils. This time, her eyes had spotted a few bone fragments weathering down the slope of a small hill. David Loope sighted the other skeleton on one of his forays in search of geological clues that could shed more light on the ancient climatic conditions.

This was exciting news: our quest to find adult skeletons inter-mingled with the eggs had finally proved successful. These titanosaur skeletons were less than ten feet away from nests, and the one found by Andrea showed evidence of being scavenged. Its bones had been broken by the sharp teeth of predatory dinosaurs, which were mixed in with them. It seems likely that the carcass of this titanosaur had remained exposed on the surface of the ancient floodplain while hundreds of females were laying their eggs. Whether this animal and the other sauropods we found in the egg layers were part of the nesting community of Auca Mahuevo is unknown. But that they all belong to the group of sauropods known as titanosaurs, the most abundant dinosaurs of the late Cretaceous in Patagonia, lends some weight to the suggestion that the eggs were laid by titanosaurs, too.

In the midst of these days of giddy discoveries, disaster nearly struck in the form of a massive electrical storm. At the quarry, a lightning bolt struck dangerously close to our team. Some of the crew members were "blown" from their places by a thunderous bolt, and others suffered minor burns when sparks traveled through their eye-glasses and metallic pieces of their garments. This warning, needless to say, sent them scurrying to camp before the rest of our paleonto-logical troop. Luis and others, including the geologic team, were working on the eggs and sauropods in egg layer 4, several miles farther from camp. By the time the rain forced Luis's group to head back, the road had reverted to a slippery path through the ancient Cretaceous floodplain, which we unsuccessfully tried to negotiate for an hour. Any hope of reaching camp vanished quickly when we confronted a pow-erful flash flood that had completely obliterated the road. Accumu-lating the drainage from a vast portion of the foothills of the Auca Mahuida, this newborn torrent sliced through our road with four-foot-tall waves of brownish water and cascaded at high velocity into a deep creek on the other side of the road. We waited out the flood for a few hours, and when we saw the first signs that the river was lowering, we left the trucks behind and crossed it on foot. Several miles later, after fording an even wider river that ran closer to camp, we reached the safety of our tents. We were all exhausted and eager to rest, which we did for the entire next day.

On March 14, several of us went to examine a curious egg clutch that Alberto Garrido had spotted a couple of days earlier. Alberto had noticed that, unlike other clutches, this one had been laid on a sandy substrate left by an ancient river that had dried up before the dinosaurs had laid their eggs. He had also noticed that the eggs appeared to have been laid in a somewhat rounded hole in the sand, which was covered by greenish red clay. Alberto had made another exceptional find.

When we got to Alberto's site, we immediately knew that we were looking at the first-known well-preserved sauropod nest. More than twenty-five eggs had been laid in a depression surrounded by a tall, sandy rim. Close examination of this depression revealed that the cross-bedded sands initially deposited by the stream had been disrupted, and that the rim of the nest was formed by massive, structureless sand, all of which indicated that the depression had been dug by the female dinosaur who had laid the eggs. The clay covering the eggs was vivid testimony of the flood that had later inundated the area. It was also significant that this nest was not located in egg layer 3, the bed of eggs that we had explored most extensively, but rather in a stratigraphically higher layer. After a couple of days of hiking, the geological team established that the nest was in the same rock layer as egg layer 4. This confirmed our suspicions that the same kind of natural catastrophe that had buried the nesting colony of egg layer 3 was responsible for the demise of the nests in egg layer 4.

During the following days, three other examples of nest structures were found across six hundred feet of the sandy bed of this abandoned river channel. All of these nests exhibited the same layout, with eggs contained in round or more irregular bowls, about three to four feet across, surrounded by elevated rims. This evidence convinced us that all the other clutches we had found had originally been laid in depressions excavated by the females. We had not recognized this before because, in all other instances, the females had chosen the muddy substrate of the floodplain to lay their eggs on. This had prevented us from observing any differences between the clay in which the eggs had been laid and the clay in which the eggs had become buried in the flood. Fortunately for us, a handful of females used the sandy bed of an abandoned river channel to lay their eggs. Because these nests were dug in sand and covered by flood-generated clay, their original structure became detectable to the geologic eye. This find not only gave us our first glimpse of the structure of a sauropod nest but also indicated that, at the time of nesting, the Auca Mahuevo sauropods did not have a strong preference to lay their eggs on a particular substrate.

With four nests to clean, map, and analyze, in addition to the ongoing excavation projects, our working schedule became frenzied. As if this were not enough, we soon realized that the nests were too big and too fragile to be collected; sadly, we would have to leave them to the mercy of erosion, which would destroy them with the next rain. The only alternative we had was to make molds of the nests so that they could be replicated back in the lab. This would entail not only a lot of work but also a lot of molding material, of which we had almost none. After carefully considering the time we had left, we decided to concentrate our efforts on the two best-preserved nests. Luckily, Nancy Rufenacht had gained a lot of experience molding big dinosaur bones back in Wyoming—eventually she would show how worthy she was to be nicknamed the Latex Queen of New Orleans. Nancy, a skilled fossil preparator from the Natural History Museum of Los Angeles County, was making her Patagonian debut. With her strong New Orleans accent and her superb sense of humor, Nancy soon became the life of our new-millennium party. She put together a list of molding tools and materials—jars of latex, silicon and plastic resin, wooden spatulas, and other tools—which we immediately undertook to find in Neuquén.

Nancy, Adrian, Gary Takeuchi, and a few others joined forces to mold the nests. Gary now hails from the Los Angeles County Museum and soon became known as Taka. With his jovial spirit and "exotic" Japanese background, Gary quickly became the best friend of all the gauchos in the area. With his experience gained through years of collecting fossil mammals at the famous La Brea tar pits and elsewhere in southern California, Gary became an invaluable asset for our crew. Nancy supervised the operation, training the others in the art of molding specimens in the field. Working in teams of two, they had to clean, mold, and then remove the molds from the nests, while carefully monitoring weather changes, since unexpected rain or dust storms could ruin the molds. Before removing a mold, they also had to construct a sturdy covering over the soft latex mold itself, so that an undeformed cast could be made from the mold back in the lab. It would take the remaining two weeks for her and the three other crew members to mold and demold these large nests. The day the molds were removed, much of the crew had to help; hours of steady pulling on the latex were required to free the molds from the nests, but by the end of the season, we had two enormous burrito-like silicon wraps to take back home for further analysis.

While many of us were busy mapping eggs and molding nests at egg layer 4, our geologists David Loope and Jim Schmitt, who works with Frankie at Montana State University in Bozeman, were looking for other clues in the rocks to interpret the environment at the time these eggs were laid. Tall and energetic, Jim strolled all over Auca Mahuevo pointing at numerous important geological features that would clarify our view of the environment and preservational aspects of the nesting site. Specifically, David and Jim had been examining two extensive, whitish rock layers a few feet above egg layer 4. David, a man of few words, is the kind of researcher who gathers a lot of data before reaching a conclusion.

On March 16, David and Jim told us that those whitish horizons were enormous layers of dinosaur tracks, surfaces that had been stepped on by thousands of sauropod dinosaurs. To confirm this interpretation, they wanted to remove the overburden of clay across a large area and examine the whitish horizons on a fully exposed surface. On March 18, several members of our crew worked down through the overburden on a long, low hill until they reached the whitish rock layer. We saw, to our amazement, that the horizon was made of distinct bowl-shaped structures containing a white mineral. These white bowls were distinct from the reddish clay that surrounded them. With this piece of evidence, the interpretation of our geological team was fully confirmed. The whitish horizons represented wet surfaces on which the dinosaurs had walked. The large, bowl-shaped depressions formed under the weight of their feet remained exposed for years, accumulating a shallow film of water during the wet season. With the evaporation of this water, a variety of precipitates added a layer of white minerals to the bottoms of the bowls. Because these surfaces would have been exposed for many decades, the deep dinosaur tracks would have accumulated the one-to-two-inch-thick deposits of mineral precipitates that we were observing. The lateral extension of these track horizons helped us connect up some of the discontinuous exposures of egg layer 4 and gave us a better sense of the size of this immense nesting colony. This discovery, too, indicated that the area was highly frequented by sauropods.

The two titanosaur skeletons that we had found intermingled with egg clutches at egg layer 4 were collected soon after we found them on March 12. This did not require a lot of excavation because they were quite incomplete. One of them lay exposed on the surface, only a foot and a portion of its tail preserved, so it took only a day for Luis, Sara Bertelli, and our illustrator, Nick Frankfurt, to encase these remains in protective plaster jackets. The other specimen, which had been scavenged, was also quite incomplete, but it took Rodolfo, Andrea, and several others a bit longer because of the overlying rock that had to be removed before trenching around the bones and encasing them in plaster. To protect themselves from the blistering sun, the team constructed a large canopy over the excavation using one of our field tarps. By March 17, these two specimens had been collected and a large array of plaster blocks lay under the shade of our trailer.

Our work collecting adult skeletons from the egg layer was still far from being finished. On March 18, while Alberto and other members of our team were exploring a series of outcrops several miles north from our main egg sites, a much more complete dinosaur skeleton was found. Although we could see many bones sticking out from a low hill, we were not sure at first what kind of dinosaur this was. Nonetheless, we knew that it was important: it was well preserved and apparently quite complete. But most significantly, it appeared to be lying on rock from egg layer 3, even though the eggs were miles to the south.

Using stratigraphic correlation, we determined that this specimen was indeed in egg layer 3. We did not have any remnant of egg layer 3 at this spot, but egg layer 4 was exposed several feet above this section. Alberto measured the thickness of rock between egg layer 4 and the layer containing the new skeleton. Then, comparing it with the thickness of rock separating egg layers 3 and 4 at our main egg sites, he discovered that the new skeleton was exactly the same distance below layer 4 as egg layer 3. In other words, when this dinosaur had died, others were laying eggs only a few miles away.

Although we did not know how much of this skeleton was still there, it would clearly take a lot of work and time to collect all the bones. The first thing we had to do was expose as many bones as pos-

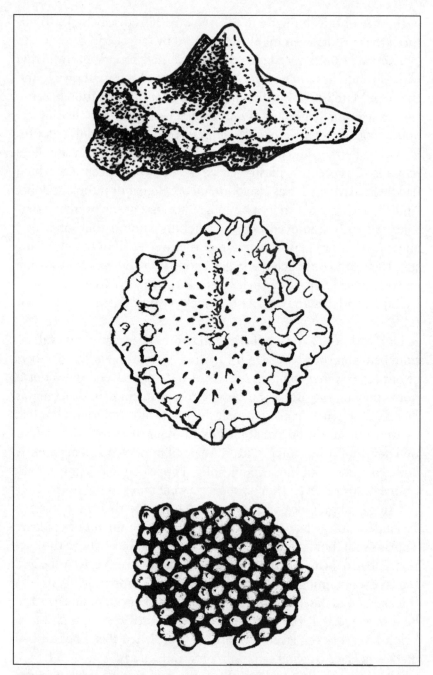

The body of the Argentine titanosaur *Saltasaurus* and those of some of its relatives were protected by a robust armor of large (top and center) and small (bottom) bony scutes.

sible by removing the overburden from the low hill. Rodolfo and a team of our paleontologists volunteered for this backbreaking job. A few days later, several tail vertebrae, ribs, a few bones from the hind limb and pelvis, and several other bones were exposed, and with this anatomical information, we were able to identify the new skeleton as that of a titanosaur sauropod.

This was another interesting coincidence in which a titanosaur was preserved in the same rock layer as the eggs, so we were inclined to suspect that the same type of dinosaur had laid the eggs. In addition to providing this important stratigraphic clue, the new skeleton was much more complete than any other titanosaur previously found in the Río Colorado Formation, the regional rock layer containing all the eggs from Auca Mahuevo. As previously noted, titanosaurs lived for at least 80 million years from the time they originated in the late Jurassic to the time they went extinct at the end of the Cretaceous. During this period, several groups evolved and differentiated into numerous species.

Neuquensaurus, the titanosaur previously known from the Río Colorado Formation, is a member of the saltasaurines, a group of titanosaurs that includes the armored *Saltasaurus* from the late Cretaceous of northern Argentina. Like *Saltasaurus,* the skin of *Neuquensaurus* and other saltasaurines was studded with bony scutes of various sizes. Saltasaurines also exhibit other features in their skeletons that did not match the bones from egg layer 3. This added a new dimension to our discovery, because if the bones were not from a saltasaurine, they would be the first known nonsaltasaurine from the dinosaur-rich deposits of the Río Colorado. If the eggs were laid by nonsaltasaurine titanosaurs, this would suggest that the embryos may not have developed armor when fully grown because bony scutes are known only for saltasaurines. If so, their skin would have been naked, like that of other sauropods.

Unfortunately, the amount of work required to collect this specimen was beyond our capability, given the time we had left. Even though Rodolfo and several others labored to excavate a portion of this beast for the rest of our season, we would have needed another two weeks or so to finish excavating all the bones. Ultimately, we decided to protect the specimen thoroughly with coats of glue and leave it for future collecting. We figured we would have more time the next season, and it is always good to start an expedition knowing that you already have got something valuable to collect. Our questions concerning the identity of this specimen would have to wait. Before we left this site, we covered it with soft sand to camouflage the area and prevent looters from disturbing our discovery.

In recent years, fossil poachers have become a serious problem for professional paleontologists. Looters are seduced by the high prices that fossils command in international markets and in auction houses. Although most countries have legislation against fossil collecting with commercial interest, many fail to provide the strict and widespread vigilance needed to prevent fossil sites from being plundered. Unfortunately, Argentina is no exception, and early this season we had discovered evidence of egg looting. We knew that the publicity generated by the significance of our discovery would put the integrity of the site at risk; sadly, that day had come. We discovered several large excavations on egg layer 3, some of which had been dug a few months earlier and others very recently. The technique used for these excavations indicated that the poachers were not professionals and that by the time they had collected a few eggs they had probably destroyed several egg clutches.

Incredibly, at least one of the poachers had left a clue to his identity at the site. As Luis was inspecting some of the holes that had been excavated, he noticed some chunks of plaster scattered at the base of the hill, and when he clambered down to inspect them, he spied a small piece of paper stuck in a nearby bush. Thinking it might be a receipt that had accidentally fallen out of his pocket, he retrieved it, but upon closer examination, it turned out to be a receipt for plaster bought at a hardware store in Neuquén about a week earlier. The poacher had even signed the receipt. Amazed at this oversight, Luis gave the receipt to Rodolfo, who turned it over to the police.

Nonetheless, our spectacular site had been plundered, and it is still in jeopardy now. We see a clear and present danger for the scientific integrity of this spectacular site, which highlights our need for additional expeditions to recover as much information and material as possible before poachers destroy it.

Our days passed as some people were molding the well-preserved nests of egg layer 4, while others were exposing the titanosaur skeleton in egg layer 3 or mapping the egg clutches exposed on the flat surfaces of egg layer 4. Busy days like these pass quickly, and we needed some rest and relaxation. Luckily, our hosts at the village had something in mind.

The morning of March 25 started early, with whinnies, sounds of bells, dogs barking, and a lot of other commotion. We soon learned that more than twenty gauchos were expected to be on hand to brand and break a hundred horses. We figured that a good way of resting would be to watch an Argentine rodeo, and indeed it was. The morning activities started with branding horses; gauchos would work in teams of five or six, each on foot with a lasso and assisted by a horseman. They would select a colt or a young horse, lasso it, and mark its thigh with an incandescent iron, also managing to drink a lot of wine and beer during their breaks under the blistering sun. By lunchtime, most of the gauchos were drunk, but they kept working nevertheless. Lunch consisted of a roasted horse, plus more wine and beer. In the afternoon, with our bellies full of horsemeat and drinks, we sat in the shade to watch the continuing gaucho games. Occasionally, some courageous gauchos would attempt to break the wild horses, almost inevitably ending up on the ground. Toward evening, we headed back to our camp to prepare dinner and take care of our domestic chores. Even after dark, however, we kept hearing the gauchos' laughter and a pitched neigh or two.

This wonderful feast was a kind of farewell party. In the days that followed we started wrapping up our tasks and getting ready to return home. The wind blew mercilessly on March 26, but short of time, we persevered. The four-week season had provided us with a lot of new information. We had discovered the first adult dinosaurs in the egg layers, and we thought that some of them might represent species previously unknown to science. Many more eggs and egg clutches had been mapped. More embryos had been collected. The first nest structures of sauropod dinosaurs had been found. We had confirmed our initial thoughts about the ancient climatic conditions of the nesting site. And we had garnered glimpses of a colossal meat-eating dinosaur that would have dwarfed the already fearsome *Aucasaurus*. It had been a successful expedition, and we were sad to go. Yet we knew that a host of analyses and important discoveries awaited us at home, and that we would certainly return.

As we left, clouds surrounded the ancient volcano that for three seasons had watched quietly over our endeavors. Deep in our thoughts, we sensed a special connection, as if Auca Mahuida had benevolently chosen us to discover and investigate its most precious treasure, one that, through the interdisciplinary efforts of our team, had shed significant light on some dark mysteries of dinosaur behavior.

A Threatened Window
on the Ancient Past

Nature has preserved a priceless window on the past for us at Auca Mahuevo. As a result of our crew's discoveries and investigations, it has become possible to peer through that window and envision some of what life was like 80 million years ago on the plains of Patagonia. The vision that has emerged is breathtaking, and if it were possible for us to drop in for a visit, it would be difficult for us to recognize where we were.

Huge herds of lumbering sauropods more than forty feet long roamed the gently sloping floodplains in search of vegetation to eat, as South America drifted lazily to the west away from Africa, which lay just over the horizon to the east. At some time during the year, hundreds of females congregated in the flood basins and abandoned stream channels of the Auca Mahuevo nesting colony to dig their nests and lay their eggs. Using their enormous feet, each female scooped out a basin about three to four feet across in the dry mud of the flood basin or the sand of the abandoned streambeds. The nests were spaced five to ten feet apart, and each female laid between fifteen and forty eggs in her nest before she retreated to the periphery of the nesting colony to stand guard or wandered off across the plains in search of food. The parents probably did not stay near the nest to provide much care for their young, either while the eggs were incubating or after the embryos hatched.

Most of the year, the climate at Auca Mahuevo was dry, and the Patagonian plains baked under the relentless Cretaceous sun. During most nesting seasons, the sunlight warmed the eggs in the nests, pro-

New evidence is documenting that large carnivorous dinosaurs may have formed packs, perhaps for the purpose of hunting. The abelisaur *Aucasaurus* could have used this technique to kill the adult sauropods that nested at Auca Mahuevo.

viding the developing embryos with excellent conditions for incubation. Inside the eggs, the embryos grew to about a foot in length. Though their heads, even with their relatively large eyes, were only a couple of inches long, numerous pencil-shaped teeth had already erupted from their jaws before they hatched, and the unhatched babies exercised their jaw muscles, which they would soon need for feeding on vegetation. Their skin was reminiscent of their other reptilian relatives that lived alongside them; scales covered their delicate bodies and formed roselike and linear patterns.

However, the nesting season did not always proceed according to plan. Occasionally, storms large enough to generate floods in the shallow streams swept across the ancient plains and surrounding regions. The water in the channels overflowed the stream banks, carrying suspended particles of silt and clay that were deposited like a muddy blanket across the flood basin. On at least five occasions, this muddy blanket was thick enough to bury the incubating eggs in the nesting colony, killing the embryos inside and beginning the geologic processes that led to their fossilization.

But most years, when floods did not ravage the colony, the embryos' development inside the eggs proceeded uninterrupted, and many of the embryos hatched to begin life no longer than a baby crocodile. Over the next few decades, the survivors grew to be more than forty feet in length and to weigh several tons, making them some of the largest animals ever to walk on earth.

Floods were not the only hazard that threatened a sauropod's survival at Auca Mahuevo. At least two perilous types of predators lurked on the plains adjacent to the nesting colony. The one that we know more about was *Aucasaurus*, a twenty-foot-long relative of the fearsome *Carnotaurus*. This gracile carnivore moved swiftly on its two powerful hind legs. Although its arms were short and were probably not formidable weapons, its jaws were studded with dozens of serrated teeth for slashing flesh, and its hind feet were equipped with razor-sharp claws for taking down prey. As if these attributes were not menacing enough, its skull was adorned with small horns over the eyes. Clearly, solitary adult aucasaurs were capable of preying on hatchlings and juvenile sauropods that frequented the floodplain, and if the aucasaurs joined together to hunt in packs, even adult sauropods could have been in jeopardy.

But *Aucasaurus* was not the top predator on the ancient Patagonian plains. Based on our current but incomplete knowledge, that
distinction fell to an even larger theropod. To date, we have found
only a few bits and pieces of its skeleton, but these fragments suggest that the animal was as large as the largest carnivorous dinosaurs
yet discovered, such as *Tyrannosaurus* and *Giganotosaurus*. A solitary
predator of this size would have constituted a threat even to an adult
sauropod of the size that lived at Auca Mahuevo, and evidence exists
to suggest that some of these superpredators congregated in packs,
making them a danger for even the largest sauropods.

In all, this portrait of life in ancient Patagonia, painted in the picturesque rocks and fossils that form the modern desert landscape, has
greatly augmented our scientific knowledge about dinosaurs and the
environment they lived in. Through the window at Auca Mahuevo, we
have caught our first glimpse of what sauropods looked like when they
first hatched. We can now be certain that they did lay large eggs and
that, at least in the case of these South American sauropods, they laid
those eggs in well-developed nests contained within a massive nesting
colony frequented by at least hundreds of mothers at one time. The
rocky pages of evolutionary history at Auca Mahuevo document that
these sauropods returned to the nesting colony numerous times to
breed and thereby sustain their majestic evolutionary lineage.

These discoveries and the insights that they have provided represent a true paleontologic treasure. Yet the clues buried in the rocks
required to reconstruct the ancient events at Auca Mahuevo were not
easily found. Serendipity played its usual role, but beyond that, the discoveries represent determined and well-coordinated scientific
sleuthing by all of the crew members who participated in our expeditions. As we have documented, such work does not simply involve
the ecstasy of initial discovery. To see clearly through the window on
the past that Auca Mahuevo provides, that initial thrill must subsequently be supplemented with laborious and meticulous scientific
investigations, both in the field and back at the lab. That work
requires the expertise of dozens of geologic and paleontologic specialists, as well as a lot of time and money, and we are extremely grateful to all of our colleagues who have lent their knowledge and to all of
the organizations that have financially supported the research.

The most rewarding part of our work is that, although this book is nearing its conclusion, the story of Auca Mahuevo is just beginning to be revealed. Although three full-scale expeditions have been mounted and more than three years of research have been conducted, numerous questions still remain. Exactly which kind of sauropod laid the eggs? Were they laid by titanosaurs, as we suspect, or by another kind of sauropod? If titanosaurs did lay the eggs, which of the many known species was responsible? Also, exactly how did the eggs and embryos become fossilized? We are sure that floods buried the eggs and nests in mud, but what processes of mineralization operated quickly enough that the poorly formed embryonic bones and skin became fossilized before they could decay?

Other mysteries concern the predators that lived on the ancient floodplain. For example, what really killed the *Aucasaurus* individual that we discovered? What kind of dinosaur did the huge, isolated theropod bones that we found belong to? Does their presence indicate that *Giganotosaurus* survived for millions of years longer than previously thought and preyed on the sauropods of Auca Mahuevo, or was a previously unknown predator of gigantic proportions roaming the plains?

We would also like to learn more details about the environment that existed at Auca Mahuevo. What kind of plants were the giant sauropods eating? Our attempts to recover fossilized pollen have thus far proved unsuccessful, and although we have found fossil stems of horsetails and wooden branches, we have yet to find rock layers that preserve abundant fossil plants. Such discoveries could go a long way toward providing clues about what vegetation lived at the site and what the sauropods ate.

Several geologic jobs are also still unfinished. One important task is to map where the egg layers and other distinctive rock layers are exposed on the surface of the ground, which would require high-resolution photographs to be taken from a low-flying airplane equipped with a specialized camera. We would then be able to map where the egg layers are found on the photographs. In addition, we have yet to find a layer of ancient volcanic ash near the site that can be used to obtain an age for the fossils through radioactive techniques. We think we know where one might be found about fifty miles west of Auca Mahuevo; however, we have not been able to get to that spot because of its lack of roads and the rough terrain.

The paleontological and geological research required to paint a complete portrait of the animals and environment that existed at Auca Mahuevo has just begun. Our view through the window to the past is still murky, and our efforts to see more clearly will no doubt require the participation of more geologic and paleontologic specialists. But over the coming years, as new expeditions and analyses are conducted, we anticipate that Auca Mahuevo will once again prove magical and reveal more precious secrets and treasures from the history of life for us to ponder and investigate.

For more discoveries to be made, however, the site must be protected from poaching and vandalism. In the past decade, with fossils fetching handsome prices at auctions and curiosity shops, paleontologists have seen their sites looted and excavated, resulting in a great loss of information. Fossils provide unique clues to reconstruct the history of ancient life, but as this book illustrates, many of those clues reside in the geological context of the fossil. Without this accompanying data—such as the exact rock layer in which it was found, the type of rock containing it, the precise location of the site, the associated fossil biota, and the environmental setting—a fossil loses much of its scientific value. This information must be preserved for posterity so that future generations of paleontologists can reinterpret the fossils in light of subsequent discoveries. Museums and their paleontologists not only have the training to collect both the fossils and their contextual information but also the infrastructure to protect that information for the future. These institutions make the fossils and their supporting data available to outside researchers and the public as well. Amateurs and commercial collectors, on the other hand, often recover minimal information about the geological and paleontological context of a fossil when they collect it, focusing only on the intrinsic value of the fossil. Furthermore, fossils recovered by amateurs and commercial collectors are often left outside the custodial care of an institution that can protect them for future generations, making them unavailable to researchers and the public.

As we mentioned earlier, we found disturbing evidence during our 2000 expedition that commercial or amateur fossil collectors had visited the site and destroyed some nests to collect eggs. If such activities become more widespread across the site, the important fossils and scientific evidence needed to answer the remaining mysteries about the site will be destroyed. As a result, knowledge about the magnificent animals that lived at Auca Mahuevo 80 million years ago could be forever lost, and our priceless view through Auca Mahuevo's window on the past could be forever clouded.

Fortunately, some initial steps have already been taken to protect this unique paleontological resource. Led by the efforts of our colleague Rodolfo, the government of the province of Neuquén in Patagonia is purchasing all the land containing the site and declaring it a fossil preserve. Eventually, plans call for a small field laboratory to be built at the site, which will be staffed by a government ranger to patrol the area and help carry on the research. Through efforts like this, paleontologists around the world have protected important fossil sites, unique repositories for our own natural heritage. Examples include Dinosaur Provincial Park in Alberta, Canada, which was designated a World Heritage Site by the United Nations, and Egg Mountain in Montana, which is owned by the Nature Conservancy. We hope that, through their being declared a fossil preserve, the rugged badlands of Auca Mahuevo will continue to yield new discoveries of fossils and related geologic insights for decades to come, and that these new fossil treasures will be available for all to see and ponder. Only in this way can Auca Mahuevo be preserved to inspire the next generation of paleontologists to continue our explorations and fill the gaps in our knowledge about past life on earth.

FURTHER READING

Carpenter, K. *Dinosaur Eggs, Nests, and Baby Dinosaurs*. Bloomington: Indiana University Press, 1999.

Carpenter, K., K. Hirsch, and J. Horner. *Dinosaur Eggs and Babies*. Cambridge, Mass.: Cambridge University Press, 1994.

Chiappe, L. M. *Dinosaur Embryos: Unscrambling the Past in Patagonia. National Geographic* 194, no. 6 (1998): 34–41.

Currie, P., and K. Padian. *The Encyclopedia of Dinosaurs*. San Diego: Academic Press, 1997.

Dingus, L., and L. M. Chiappe. *The Tiniest Giants*. New York: Random House, 1999.

Dingus, L., and T. Rowe. *The Mistaken Extinction*. New York: William H. Freeman, 1997.

Gillette, D., and M. Hallett. *Seismosaurus, the Earth Shaker*. New York: Columbia University Press, 1994.

Horner, J., and E. Dobb. *Dinosaur Lives: Unearthing an Evolutionary Saga*. Orlando: Harcourt Brace, 1998.

Norell, M., and L. Dingus. *A Nest of Dinosaurs: The Story of Oviraptor*. New York: Random House, 1999.

Norell, M., E. Gaffney, and L. Dingus. *Discovering Dinosaurs*. New York: Alfred A. Knopf, 1995.

Novacek, M. *Dinosaurs of the Flaming Cliffs*. New York: Anchor, 1997.

Sloan, C. *Feathered Dinosaurs*. National Geographic Society, 2000.

INDEX

Page numbers in *italics* refer to illustrations.

Illustration Credits

page 30: Glyptodonts and giant sloths. Illustration by Charles R. Knight, courtesy of American Museum of Natural History, New York.

page 32: Geologic and evolutionary time scale. Illustration by Portia Rollings, from Lowell Dingus and Luis M. Chiappe, *Tiniest Giants.* (New York: Random House, 1999.)

page 61: Sprawling and erect postures. Illustration by Frank Ippolito.

page 62: Genealogical relationships of dinosaurs. Illustration by Timothy Rowe, from Lowell Dingus and Timothy Rowe, *The Mistaken Extinction.* (New York: W. H. Freeman & Co, 1997.)

page 66: Genealogical relationships of theropods. Illustration by Timothy Rowe, from Lowell Dingus and Timothy Rowe, *The Mistaken Extinction.* (New York: W. H. Freeman & Co, 1997.)

page 108: Map of the world 80 million years ago. Illustration by Frank Ippolito.

page 113: Magnetic polarity of Auca Mahuevo. Illustration by Portia Rollings, from Lowell Dingus and Luis M. Chiappe, *Tiniest Giants.* (New York: Random House, 1999.)

page 120: Floodplain environment. Illustration by Frank Ippolito.

Photographs following page 160

All photographs by L. Chiappe, with the exceptions of no. 8, Embryonic skin, by L. Meeker, and nos. 10 and 11, Skull and limb bones of an unhatched sauropod and Nearly complete embryonic skull, by D. Meir.

Luis M. Chiappe, a 1996 Guggenheim fellow, is Associate Curator and Chairman of the Department of Vertebrate Paleontology at the Natural History Museum of Los Angeles County. Dr. Chiappe has conducted extensive fieldwork in the Americas as well as Asia. His many articles have appeared in *Nature, Science, National Geographic, Scientific American,* and *Natural History.* He lives in Santa Monica, California.

Lowell Dingus is a research associate in the department of paleontology at the American Museum of Natural History in New York City and the president of InfoQuest, a private, nonprofit foundation devoted to public education and research. The coauthor of *Discovering Dinosaurs, The Mistaken Extinction,* and several children's science books, he lives in New York City.